Transition in Energy Sector: Scope, Challenges and Future Opportunities

Transition in Energy Sector: Scope, Challenges and Future Opportunities

Edited by
Ayan Banik

CWP
Central West Publishing

This edition has been published by Central West Publishing, Australia

For more information about the books published by Central West Publishing, please visit https://centralwestpublishing.com

Disclaimer
Every effort has been made by the publisher, editors and authors while preparing this book, however, no warranties are made regarding the accuracy and completeness of the content. The publisher, editors and authors disclaim without any limitation all warranties as well as any implied warranties about sales, along with fitness of the content for a particular purpose. Citation of any website and other information sources does not mean any endorsement from the publisher, editors and authors. For ascertaining the suitability of the contents contained herein for a particular lab or commercial use, consultation with the subject expert is needed. In addition, while using the information and methods contained herein, the practitioners and researchers need to be mindful for their own safety, along with the safety of others, including the professional parties and premises for whom they have professional responsibility. To the fullest extent of law, the publisher, editors and authors are not liable in all circumstances (special, incidental, and consequential) for any injury and/or damage to persons and property, along with any potential loss of profit and other commercial damages due to the use of any methods, products, guidelines, procedures contained in the material herein.

NATIONAL LIBRARY OF AUSTRALIA

A catalogue record for this book is available from the National Library of Australia

ISBN (print): 978-1-922617-28-6

About the Editor

Ayan Banik obtained his bachelor's degree in engineering from Cooch Behar Government Engineering College, India and M. Tech. from the National Institute of Technical Teachers' Training & Research, Kolkata, India. He has received C. Engg. from the Institution of Engineers (India) in 2021. He has authored/co-authored several research papers and monographs in peer-reviewed journals & conference proceedings indexed by Scopus/Web of Science (WoS). He has contributed to many national and international patents in engineering and technology sectors. In addition, he has been the recipient of numerous prestigious awards for his dynamic scientific activities and outstanding research contribution. He has also been an integral part of notable scientific and professional societies, i.e., IET, IEEE, IEI, and ISCA. His field of interest includes electric power quality, smart grid, fault diagnosis, power system protection, signal analysis, DERs, energy management, and waste-to-energy technologies.

Contents

Preface

Renewable energy today only accounts for a small fraction of the primary energy used by the world. Cost and the intermittent nature of most of these sources have been the main obstacle to increasing the use of renewable energy. This book reviews the current status of wind, solar, biomass, hydroelectric, wave, tidal, and geothermal energy sources, including estimates of the available supply of each and current prices. As existing solar photovoltaic cells can now provide energy at prices close to that of other sources, it can be expected that solar energy will witness a rapid increase in use parallel to the increase in wind energy over the past decade.

This book provides an encyclopedic forum for disseminating and exchanging up-to-date scientific information on theoretical, generic, and applied areas of knowledge. The topics deal with new devices and circuits for cutting edge energy systems, photovoltaic/solar, thermal, wind energy systems, tidal energy, fuel cell systems, bioenergy, geoenergy, sustainable energy resources, energy storage systems, energy market management-economics, off-grid isolated energy systems, energy in transportation systems, energy resources for portable electronics, intelligent energy power transmission, distribution & interconnectors, energy-efficient utilization, environmental issues, energy harvesting, nanotechnology in energy, policy issues on renewable energy, building design, power electronics in energy conversion, new materials for energy resources and RF powered magnetic field energy devices.

The chapters authors have made an effort to accommodate all forms of literature, analytical examples, and most importantly, present status quo, which shall benefit the reader to understand the topic better and trigger their capabilities to contribute significant innovation in the near future to have a better future sustainable global energy.

1

Sustainable Energy Generation from Biomass Gasification: Scope and Challenges

Bheru Lal Salvi[1] and Ritu Raj Jaitawat[2]

Department of Mechanical Engineering, College of Technology and Engineering, Maharana Pratap University of Agriculture and Technology, Udaipur – 313001 Rajasthan, India

Abstract

In an era of energy crisis and increasing environmental pollution, energy from biomass may play a very crucial role in the global energy system. The energy from biomass, popularly known as bioenergy, is a very flexible source that can be turned up and down to boost energy supplies in nations with rising demand. This chapter deals with introduction to biomass and its conversion methods for value added fuel development. Majorly, the biomass gasification technology is discussed in detail. Gasification converts biomass into combustive syngas that ideally contains all the energy originally present in the biomass. Cleaning and purification of syngas can improve its efficiency as a fuel. Considering the rising petroleum prices and increasing contributions made by fossil fuels to global warming, biofuels can be predominantly used in compression ignition (CI) engines for transportation purposes and power generation applications. Apart from the conversion of biomass to biosyngas, byproducts of biomass gasification can also be used as lubricants, industrial chemicals and many other products currently derived from petroleum or natural gas. However, there are downsides too for bioenergy technology. Various limitations hindering the wide range application of biomass as renewable source or energy are discussed in this chapter. Biofuel production requires growing energy crops which leads to the clearing of forests and a large amount of energy is consumed in the production and conversion processes of biofuels. Thus, proper sourcing and processing of biomass for energy are required to make bioenergy a sustainable source of renewable energy.

1.1 Introduction

Energy is one of the key inputs and resources for the social and economic development of any country. The overall global energy demand is increasing regardless of the type of energy source. It is predicted that the world energy demand will experience an increase of 19% by the year 2040. Currently, about 80% of energy needs are met by fossil fuels (World Energy Needs, 2021), and the remaining is obtained from energy sources such as biofuels, nuclear power, hydro and other renewable sources such as wind, solar, biomass, geothermal, etc. The type of energy source used has a major impact on environmental health. The over-utilization of fossil fuels is causing global warming, which is leading to drastic climate change (Jindal and Salvi, 2012). Hence, for sustainable economic growth, the use of renewable energy sources is the inevitable choice.

Renewable energy sources obtain energy from natural resources that can replenish themselves in lesser time than a human lifetime. These resources do not deplete the planet's resources and are available in one or the other form everywhere. Most importantly, they cause very little damage to the environment as compared to fossil fuels. These resources include the sunlight, biomass, wind, tides, waves, rain and thermal energy stored in the earth's crust. Even renewable sources have an impact on the environment, and each source has its trade-offs. However, their advantages over the disturbing impacts of fossil fuels are undeniable. There are several renewable sources available, with each offering its advantages and limitations depending on weather and time dependency, geographical location, requirements for energy generation, etc. Solar power is one the major renewable sources of clean energy, i.e., without any harmful emissions, with a huge potential for power generation. However, it is dependent on the weather, which limits its usability. Similarly, wind power is a significant source, but requires a large land area and is also weather-dependent. Biomass, on the other hand, is one of the potential sources of energy that is not dependent on weather and is easily available throughout the year. It is obtained by using organic material from plants and animals, including crops, trees and waste wood.

The energy obtained from biomass, popularly known as bioenergy, has been used by the human beings since our ancestor apes learned to use fire to cook. Since then, bioenergy was mainly used for household cooking, until people realized that efficient use of biomass can replace fossil fuels for power generation. Bioenergy can be derived from recently living organic matter known as biomass, thus, making a great backup for weather-dependent renewable technologies such as wind and solar. Biomass can be either burned directly to produce heat or converted into oil or gas substitutes. The direct combustion of biomass releases smoke, fumes, dust, volatile compounds and toxic gases because of incomplete reactions. In contrast, biomass gasification can keep down the emission of pollutants. In this regard, biomass gasification is considered an effective pathway for bioenergy generation. In this chapter, biomass availability and various methods of bioenergy generation are discussed. A major focus is provided to the gasification process for power generation and its applications.

1.2 Biomass as a Renewable Energy Source

Biomass is any organic matter that comes from living or once-living organisms such as plants and animals. In terms of energy, biomass refers to the organic materials that can be used as combustible material or energy sources for power generation. Most commonly, plants, wood and agricultural and food waste are used as biomass. These are called biomass feedstock. The plants convert carbon dioxide and water into carbohydrates in the presence of the sun and store this energy in the form of chemical energy. This energy can be extracted in several ways for power generation. The feedstocks can be grown specifically for energy generation, called energy crops, or can be collected in the form of waste products from agriculture, food processing and timber production.

Biomass is a versatile as well as a flexible renewable energy source, which can be turned up and down to meet the demand. Using various biomass to bioenergy conversion technologies, carbon from biomass and waste can be utilized to produce clean transportation fuels, renewable power and useful by-products. The biofuels can be obtained from biomass which have properties comparable to that of fossil fuels.

1.2.1 Availability of Biomass

Bioenergy is the largest source of renewable energy. Biomass can either be grown specifically for energy generation, called energy crops, or can be collected in the form of waste products from agriculture, forests, municipal and industrial waste, etc. The agricultural sector contributes less than 10% of the total biomass supply, whereas forestry products account for about 85% of the total biomass used for energy purposes. The municipal and industrial wastes, on the other hand, are majorly used for energy production in urban areas (Global Bioenergy Statistics, 2019).

World Bioenergy Association (WBA) estimated that by the year 2035, energy crops can be dedicatedly grown on 5% of the agricultural area, provided the available land has to be protected and well managed under sustainable governance. The energy production from this part of the agricultural area is estimated to be 26-34 EJ per year accounting for 43% of the total bioenergy production (Global biomass potential towards 2035, 2016). Crops such as sugarcane, sugar beets, corn, canola, cereals, jatropha, oil palm, sorghum, soyabean, cassava, etc., can be used for both food and bioenergy. While crops like energy grasses, miscanthus, switchgrass and short rotation coppices can only be used for energy generation or material purposes. In addition to energy crops, agricultural residue is one of the most promising components of bioenergy production. Agricultural residues can be classified majorly as herbaceous, woody and other forms of byproducts. The herbaceous residues include straw from rice, corn, cereals, bagasse, empty fruit bunch from oil palm, empty corn cobs, etc., while woody biomass is produced as a consequence of pruning and regenerating orchards, vineyards, olive plantations and oil palm plantations. Other byproducts include all kinds of processing residues such as sunflower shells, kernels, rice husks, animal manure, etc. WBA estimated that due to economic, ecological, market and institutional reasons, the usable energy from these agricultural residues will only be a small part of their potential. Systematic use of these byproducts may offer a potential of 30-38 EJ by 2035 (Global biomass potential towards 2035, 2016).

The forestry sector is the largest contributor to the bioenergy mix globally. The primary product from forests used for bioenergy generation is wood fuel. Wood fuel is majorly used for traditional cooking and heating purposes. However, it is estimated that by 2035, the contribution of forests for bioenergy generation will reduce from 85% to 52% viewing the rapid decline in the forest area. The third and final potential biomass is the municipal and industrial waste, which is estimated to produce 6-10 EJ of energy by 2035, contributing 5% of bioenergy generation (Global biomass potential towards 2035, 2016).

1.2.2 Classification of Biomass

Biomass is naturally available in almost every part of the world. There are numerous varieties of biomass having substantially different chemical compositions. So, there is no specific way to classify biomass. As per the vegetation type or ecology, the biomass can be classified as wood and woody biomass, herbaceous biomass, aquatic biomass, animal and human waste biomass, biomass mixtures, etc. (Tursi, 2019).

Wood and Woody Biomass

Woody biomass is a plant material that consists of carbohydrates and lignin. It is mainly obtained from trees and shrubs and can include roots, trunks, bark, branches, leaves, limbs and vines. Woody biomass can be collected from many sources such as otherwise non-merchantable trees, residues from forestry operations and timber industries, wildfire fuel-reduction thinning, trees, shrub cuttings from urban area maintenance process, sawmill residues, natural forest and trees grown for energy. Woody biomass is a potential source of renewable energy and is expected to satisfy about 18% of the worlds' primary energy needs in 2050 (Lauri et al., 2014).

Herbaceous Biomass

New European Pellets Standard EN 14961-1 defines herbaceous biomass as the non-woody stem of plants that dies back at the end of the growing season. Such biomass includes grains and their by-products such as cereal straw (Alakangas, Eija, VTT, New European Pellet Standard EN 14961-1).

Herbaceous biomass can be obtained from agricultural residues and energy crops (Tursi, 2019). Agricultural residues are either generated as the byproducts during crop harvesting or processing of crops. The residues obtained from crop harvesting are called the primary or field residues, while the ones collected during crop processing are called secondary or processed based residues. The field residues are sometimes used as fertilizers, fodder for animals, or for controlling erosion, or are simply burned before the start of the next growing season. On the other hand, the processed residues have a high potential for energy generation and are a good energy source as these do not compete with food production (Mohammed et al., 2018). Global monitoring of agricultural residues is difficult, and their bioenergy potential is also not very clear. The energy crops are specifically grown for energy generation, and their sustainable utilization can bring a positive impact on the global energy sector by replacing fossil fuels. At the same time, inefficient exploitation of such biomass can add to the pressure on global food demand.

Aquatic Biomass

Aquatic biomass includes photosynthetic algae (macro and micro algae) and photosynthetic cyanobacteria. They have great potential to produce biomass and lipids per hectare than terrestrial crops. Some species can directly produce fuels, such as hydrogen, ethanol, or alkanes. Microalgae are more well-known for their high growth rates and impressing lipid content, whereas macroalgae have more carbohydrate content that may be used to produce bioethanol, syngas and biogas, or simply used as biomass for combustion. As compared to terrestrial crops, the aquatic biomass has a higher growth rate and can be continuously harvested. This biomass does not compete with food crops and has many environmental benefits (Rasmussen et al., 2009).

Animal and Human Waste Biomass

Animal and human waste biomass includes bones, meat meal, various types of animal manure, human dung, etc. Undigested and partially digested dietary nutrients in manure help in plant growth and add organic matter to the soil and, hence, this biomass is mainly used as fertilizer on agricultural land (Mukhtar and Capareda, 2017). However, this waste has raised many concerns related to pollution, health, and odor, which require its proper waste management. Anaerobic digestion was introduced as a method for converting this waste into biogas. The biogas can be burned directly to produce energy for cooking and heating or utilized for electricity generation in turbines and internal combustion engines (Horan, 2018).

Biomass Mixtures

Biomass mixture includes a mixture of several substrates from different categories of biomass mentioned above. These are mainly the waste after utilization of the biomass and its derivatives.

1.2.3 Biomass Conversion Technologies

Numerous conversion processes can be used to convert biomass into a suitable form of energy. The choice of conversion process depends on the type and availability of biomass feedstock, end-use requirements, economic and environmental conditions. project and site-specific factors, etc. (McKendry, 2002b). Major conversion technologies are thermo-chemical conversions, biochemical conversions, physicochemical conversions, etc. Depending on the conversion technology used, a few physicochemical properties of biomass are significant for further processing. These properties include moisture content (intrinsic and extrinsic), caloric value, proportions of fixed carbon and volatile substances, ash content, alkali metal content and cellulose/lignin ratio. The conversion process of dry biomass is majorly influenced by the first five properties, while moisture content and cellulose/lignin ratio are essential for biochemical conversion processes (McKendry, 2002a; Prasad et al., 2015).

Thermo-chemical Conversions

Thermochemical conversion is one of the several pathways for intermediate energy carriers or heat production. It involves controlled heating and/or oxidation of biomass. According to the oxidation environment heating rate, and particle size, the thermochemical conversion methods are classified into four processes. These processes are combustion, pyrolysis, gasification and liquefaction (Tanger et al., 2013).

Combustion: Direct combustion of biomass is the principal energy pathway contributing approximately 90% of the world's total renewable energy produced from biomass. Carbon and hydrogen rich biomass is burned in the presence of oxygen at high temperatures to produce carbon dioxide, water vapor and heat (Eq. (1) and (2)) (Tursi, 2019).

$$\text{Biomass} + \text{Oxygen} \rightarrow \text{Carbon Dioxide} + \text{Water} + \text{Heat} \quad (1)$$

The approximate chemical reaction for biomass combustion with pure oxygen is given by Eq. (2).

$$CH_{1.44}O_{0.66} + 1.03O_2 \rightarrow CO_2 + 0.72\, H_2O + Heat \quad (2)$$

As shown in equations (1) and (2), the chemical energy stored in biomass is converted into heat, making combustion an exothermic chemical reaction. This chemical energy can be converted into mechanical power or electricity using various process equipment. The temperature of hot gases produced from combustion range between 800 to 1000 °C. Any type of biomass having moisture content less than 50% can be used for the combustion process. The biomass combustion plants have net bioenergy conversion efficiency between 20 to 40% (McKendry, 2002b).

Pyrolysis: Pyrolysis involves partial combustion of biomass in the absence of oxygen at a temperature of around 500 °C. This process converts biomass into liquid (bio-oil or pyrolysis oil), gaseous (combustible gas) and solid (charcoal) products (McKendry, 2002b). Vaporization of volatile components of the biomass takes place at high temperatures, leading to the production of gases. The vapors of these gases are condensed into liquids by liquefaction (Tursi, 2019). Pyrolysis oil is a heterogeneous mixture having high oxygen content and alkalinity, suitable for upgradation to fuels and chemicals (Tanger et al., 2013).

The nature and yield of pyrolysis products are determined by pyrolysis conditions such as temperature, heating rate, pressure, residence time, catalyst, environment, biomass particle size and nitrogen flow rate (Verma and Mondal, 2016). High temperature and long residence time produce non-condensable gaseous products; high temperature and short residence time form condensable fractions; and low temperature yields solid products (Bridgwater, 2012). Depending on the various pyrolysis conditions, the process is classified as:

- **Fast pyrolysis:** In fast pyrolysis, finely ground biomass feed is heated at high temperatures for a very short time to prevent further cracking of the pyrolytic product to non-condensable compounds (Salvi et al., 2021). The temperature in fast pyrolysis ranges between 450-600 °C with a very short residence time of less than two seconds. The fast pyrolysis of wood is reported to give liquid yield as high as 75% (Bridgwater et al., 2009; Hornung, 2014; Garba, 2020). The fast pyrolysis is very popular because of the need for bio-fuel oil.

- **Intermediate pyrolysis:** The operating temperature of intermediate pyrolysis ranges between 300-500 °C, resulting in less viscous liquid containing low tar. Large-sized biomass feed can be used, giving around 55% liquid yield (Hornung et al. 2011; Garba, 2020).

- **Slow pyrolysis:** In slow pyrolysis, carbonization of biomass feed takes place without condensing the pyrolysis products. It is majorly used for charcoal and bio-char production but can also be used for liquid fuels and biogas (Stamatov et al. 2006). The operating temperature of this process ranges between 200 to 400 °C with a long residence time, and a low heating rate of about 5–10 °C /min and a wide range of particle size from 5 to 50 mm can be used (Demirbas and Arin, 2002).

Pyrolysis offers a key benefit of converting solid biomass feed into liquids, gases and vapors which are cost-effective to handle, transport and store (Sharma et al., 2014).

Gasification: The gasification process involves partial oxidation of carbonaceous (hydrocarbon) material at high temperatures (around 800–900 °C) to produce synthesis gas or syngas. Syngas is mainly composed of carbon monoxide (CO), hydrogen (H_2) and nitrogen (N_2) (McKendry, 2002b). CO and H_2 together account for 85% of the total volume, and the remaining gas contains CO_2 and CH_4 (Molino et al., 2016). Biomass gasification is carried out in a closed high-pressure and high-temperature chemical reactor, called a gasifier, containing a limited amount of oxygen. As biomass flows through the reactor, it undergoes drying, pyrolysis, combustion and reduction processes (Bankhile et al., 2016). Syngas obtained is then cleaned to remove impurities such as tar, particulate matters, etc., along with traces of sulfur and ammonia. A typical gasification system consists of a gasifier, a gas cleaning and purifying system and an energy recovery system (Belgiorno et al., 2003).

Depending on the biomass type and operational conditions, the produced gas has many combustible and non-combustible fractions. In general, syngas has a calorific value of between 4.5 and 6 MJ/m^3 (Akia et al., 2014). Syngas can be used as a fuel in internal combustion engines or as a raw material for the production of second-generation biofuels such as methanol, dimethylether (DME), ethanol, Fischer Tropsch diesel, hydrogen and synthetic natural gas (SNG) (Molino et al., 2018).

Liquefaction: The liquefaction process is carried out in the liquid phase at moderate temperatures (280-370 °C) and high operating pressures ranging from 10 to 25 MPa. The main product of this process is liquid bio-granulate like crude oil. It also generates other gaseous, aqueous and solid byproducts (Barreiro et al., 2013). A chemically stable fuel with high H/C ratio oils, low oxygen content and high heating value is produced through liquefaction (Tursi, 2019).

Biochemical Conversion of Biomass

Biochemical conversion technology uses biological agents such as enzymes, bacteria, etc., to decompose biomass into available carbohydrates, which are further used to produce liquid fuels and biogas (Mahalaxmi and Williford, 2014). This technology for biomass conversion is a clean, pure and efficient process than other conversion technologies (Chen and Wang, 2016). The commonly used biochemical processes are anaerobic digestion (AD) and fermentation.

Anaerobic digestion: Anaerobic digestion involves the conversion of lignocellulosic and other forms of wastes into bioenergy. It is a cost-effective and sustainable technology and helps in proper waste management (Li et al., 2019). Commonly used feedstocks include agricultural residues, sewage sludge, municipal solid waste (MSW) and animal manure. In this process, a series of chemical reactions occur through natural metabolic pathways, enabled by microorganisms in the absence of oxygen. During the reactions, organic macromolecules are broken into simpler molecules, to produce biogas and digestate (Sharma, 2015). These nutrient-rich digestates can be used as fertilizers in agriculture (Li et al., 2019).

The anaerobic digestion is a multi-stage process with four basic steps i.e., hydrolysis, acidogenesis, acetogenesis and methanogenesis (Zamani, 2015). Hydrolysis is the first stage in which biomass is converted into smaller molecules like simple sugars, fatty acids, amino acids, etc. (Horan, 2018). In the second stage i.e., acidogenesis, further decomposition takes place in the presence of acidogenic microorganisms (fermentative bacteria). Products like CO_2, NH_3, H_2, H_2S, lighter volatile fatty acids, carbonic acids and alcohols are generated at this stage. Acetogenic microorganisms catabolize the acidogenesis products into acetic acid (CH_3COOH), H_2 and CO_2 in the third stage, i.e., acetogenesis. The final stage, methanogenesis, produces methane from the products of the third stage through hydrogenotrophic methanogenesis and/or acetoclastic methanogenesis, as represented by Eq. (3) and (4), respectively (Zamani, 2015; Tursi, 2019).

Hydrogenotrophic methanogenesis:

$$CO_2 + 4H_2 \rightarrow CH_4 + 2H_2O \tag{3}$$

Acetoclastic methanogenesis:

$$CH_3COOH \rightarrow CH_4 + CO_2 \tag{4}$$

Fermentation: Fermentation is a process to convert sugar crops (sugarcane, sugar beet, etc.) and starch crops (wheat, maize, etc.) into ethanol and CO_2 by microorganisms (mainly yeasts) under anaerobic conditions (Eq. (5)). This process can also convert lignocellulosic biomass, however, it involves more complex reactions and is still at the pre-pilot stage (McKendry, 2002b). Fermentation comprises a series of biochemical reactions to break simple sugars i.e., hexoses and pentoses. Glycerol and carboxylic acids are produced as byproducts along with ethanol and CO_2 (Eq. (6) and (7)) (Strezov, 2014).

$$Sugars \rightarrow Ethanol + CO_2 + by\text{-}products \quad (5)$$

$$C_6H_{12}O_6\ (Hexoses) \rightarrow 2C_2H_5OH + 2CO_2 \quad (6)$$

$$3C_5H_{10}O_5\ (Pentoses) \rightarrow 5C_2H_5OH + 5CO_2 \quad (7)$$

The process yield and quality is dependent on factors such as temperature, type of feedstock, pH, inoculum and fermentation time (Strezov, 2014).

Physicochemical Conversion of Biomass

The physicochemical conversion process involves the conversion of different types of animal fats and vegetable oils into biodiesel. Esterification and transesterification processes are used in this conversion technology. Rapeseed oil and sunflower oil can be used to produce first-generation biodiesel. Microbial oil, such as algal oil and waste oils (such as waste cooking oil), can be used to generate third and second-generation biodiesel, respectively (Tursi, 2019).

1.3 Biomass Gasification

Biomass gasification is a process of organic material decomposition into combustible gases (CO, H_2, CH_4 and CO_2) and lighter hydrocarbons, in the presence of any gasification agent at very high temperatures ranging from 600 to above 1000 °C. Various gasification agents that can be used are air, steam, carbon dioxide, nitrogen, oxygen, or a mixture of these gases. The produced gas has many impurities such as char, tar, ash and other minor contaminants (McKendry, 2002b; Bridgwater, 1995; Beenackers, 1999). Gasification of biomass can convert 60-90% of the energy in biomass into the energy in syngas. The syngas can be used for heat generation in industries, or as a fuel in engines to produce mechanical and electrical power, or as a raw material in the process of generation of secondary biofuels (Reed and Das, 1988).

1.3.1 Gasification Technology

The gasification process converts organic or fossil-based carbonaceous materials into CO, H_2 and CO_2 at elevated temperatures (>700 °C), without combustion in the presence of a controlled amount of oxygen or steam. Biomass gasification reaction can be simply expressed as Eq. (8) (Mai and Nguyen, 2020).

$$C_aH_bO_cN_dS_e + O_2 \rightarrow CO_2 + H_2O + N_xO_y + SO_z + Heat \quad (8)$$

This process is carried out in a chemical reactor called a gasifier. A typical biomass gasifier consists of four major compartments, i.e., drying zone, pyrolysis zone, combustion zone and reduction zone (Uppalapati.Babu and Kumararaja, 2019).

Drying zone: In the drying zone, biomass is dried at the topmost section of the gasifier using the heat from the lower parts of the gasifier (Eq. (9)).

Feedstock (moist) + Heat → Feedstock (dry) + H_2O-vapour (9)

Pyrolysis zone: The pyrolysis zone has a temperature higher than 250 °C to decompose the feedstock into volatiles and carbon (Eq. (10)).

Feedstock (dry) + Heat → Carbon (Char) + Volatiles (10)

Combustion zone: In the combustion zone, the products of the pyrolysis zone are oxidized or burned in the presence of partial air/oxygen. The combustion reactions of char and volatiles are highly exothermic, leading to an increased temperature (around 1100-1500 °C) in the zone (Chawdhury and Mahkamov, 2010; Jain, 1999). Following reactions takes place in this zone (Eq. (11) and (12)) (Reed and Das, 1988; Rajvanshi, 1986).

$$C + O_2 \rightarrow CO_2 + 393 \frac{kJ}{gmol} \quad (11)$$

$$2H_2 + O_2 \rightarrow 2H_2O + 242 \frac{MJ}{kmol} \quad (12)$$

Apart from maintaining high temperatures in the zone, these reactions oxidize almost all condensable products of pyrolysis.

Reduction zone: In the reduction zone, heat energy obtained in the combustion zone is converted into the chemical energy of the syngas produced. The reactions happening in this zone are presented by equations (13), (14), (15) and (16) (Reed and Das, 1988; Rajvanshi, 1986).

The Boudouard reaction:

$$C + CO_2 \rightleftharpoons 2CO - 172.6\frac{kJ}{gmol} \tag{13}$$

The water-gas reaction:

$$C + H_2O \rightleftharpoons CO + H_2 - 131.4\frac{kJ}{gmol} \tag{14}$$

The water-gas shift reaction:

$$CO + H_2O \rightleftharpoons CO_2 + H_2 + 42.3\frac{kJ}{gmol} \tag{15}$$

The hydro gasification reaction:

$$C + 2H_2 \rightleftharpoons CH_4 + 75\frac{kJ}{gmol} \tag{16}$$

The obtained mixture of gases is known as synthesis gas or syngas. When biomass is used as input feedstock, the syngas is known as biosyngas.

1.3.2 Classification of Gasifiers

Gasifiers can be broadly classified, based on feedstock's condition at bed, into three categories, i.e., fixed bed, fluidized bed and entrained flow. The deciding factors for the choice of gasifier are the fuels used in the gasification process and desired application of syngas.

Fixed Bed Gasifiers

Fixed bed gasifiers, also called moving bed gasifiers, have a fixed bed of solid biomass particles through which the oxidation medium (air and gas) flows in an upward or downward motion. Based on airflow direction, the fixed bed gasifiers are further classified as updraft (counter-current), downdraft (co-current) and cross-draft gasifiers. These gasifier types are best suited for small-scale power generation as they are simpler in design, but produce gas with low calorific value and high tar content (Reed and Das, 1988; McKendry, 2002c).

Updraft gasifier: In the updraft gasifier, the feedstock is fed from the top, while the gasification agent, i.e., air, steam, etc., is introduced at the bottom (Figure 1.1). The gasification agent crosses the grate and enters the combustion zone, where it helps in the combustion of the pyrolysis products. The temperature in this zone rises to around 1000 °C. The produced ash falls down through the grate, while the gas moves upward towards the reduction zone. While moving upwards, the hot gas loses heat, thereby helping in the pyrolysis and drying process of the feed coming downwards. Hence, the overall efficiency of the process is high. Due to the reduced temperature of the outlet gas (around 200-300 °C), its tar content is high, which is the major disadvantage of this process. The produced gas has low particulate content due to the filtering effect of the feed (McKendry, 2002c).

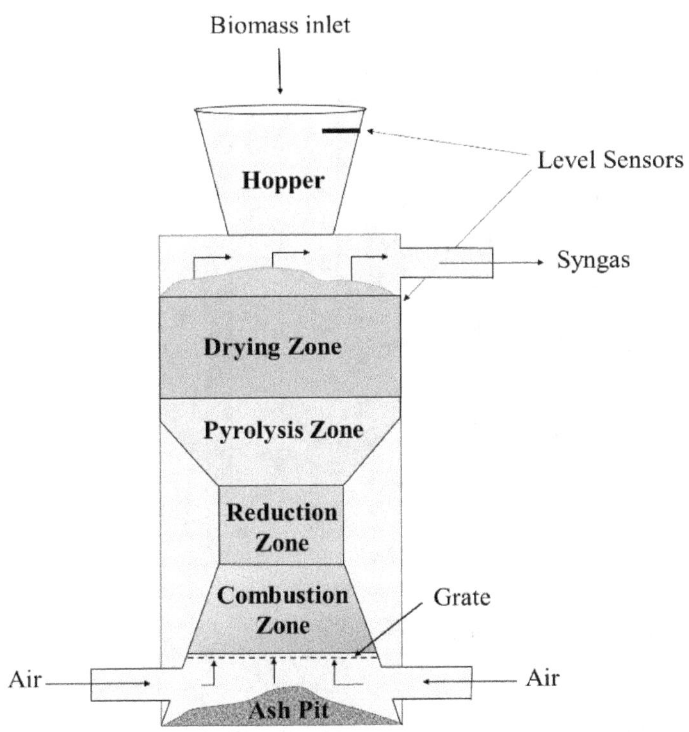

Figure 1.1 Schematic of an updraft gasifier.

Downdraft gasifier: In a downdraft gasifier, the feed is introduced from the top, while the gasification agent is fed directly into the combustion zone. So, the feed and gasification agent move in the same direction (Figure 1.2). Drying, pyrolysis, combustion and reduction are the reaction zones from top to bottom in this type of gasification unit. The produced gas leaves from the bottom at high temperatures ranging from 900-1000 °C, enabling the partial cracking of tars. Hence, the overall energy efficiency of the system is comparatively low, and the tar content in the produced gas is also low. The particulate content of this gas is higher than that obtained from an updraft gasifier (McKendry, 2002c).

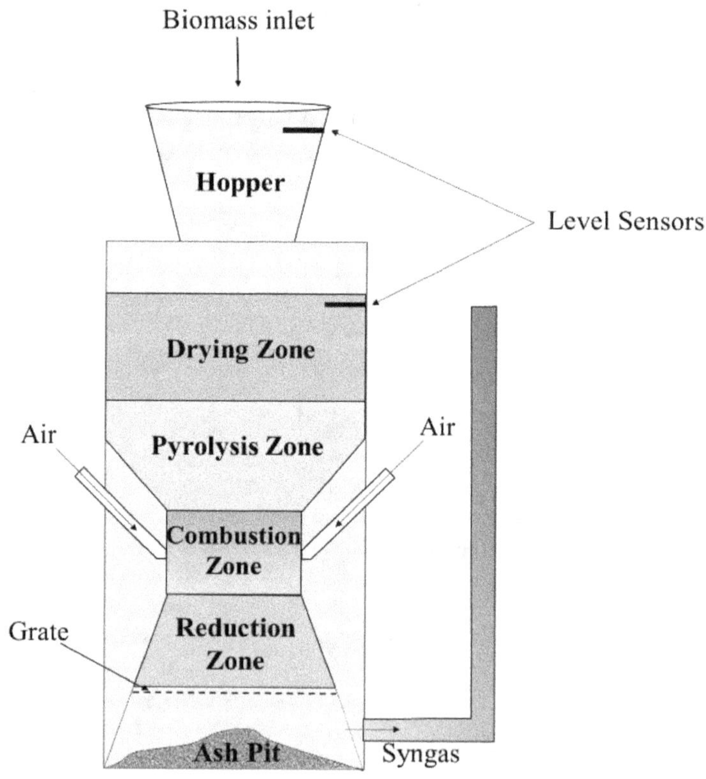

Figure 1.2 Schematic of a downdraft gasifier.

Cross-draft Gasifier: In a cross-draft gasifier, the feed is fed from the top, and air is introduced from one side of the reactor. The produced gas is withdrawn from the opposite side of the air inlet. Combustion of the pyrolysis products takes place around the air inlet, while the pyrolysis and drying process is carried higher up in the reactor. Ash falls to the bottom of the chamber, and the produced gas leaves at high temperatures (800-900 °C). Hence, the overall efficiency of the process is low (McKendry, 2002c).

Fluidized Bed Gasifier

The fluidized bed gasifier works on the principle of fluidization of a solid bed. A fluidization medium (air, steam, or air-steam mixture) is introduced in the reactor. In this reactor, the feedstock, together with an inert bed material (such as silica sand), acts as a fluid (Kunii and Levenspiel, 1991; Basu, 2006). These gasifiers can be classified as bubbling fluidized bed (BFB) gasifiers and circulating fluidized bed (CFB) gasifiers, depending on the flow velocity of the fluidization agent. In BFBs, the flow velocity of the fluidization agent is less than 1 m/s in contrast to 3-10 m/s in CFBs. Owing to the high flow velocity in CFBs, the solid particles move upward with the flow (Siedlecki et al., 2011). These particles leave the reactor with the gas and enter the cyclone separator, followed by recycling back to the bottom of the reactor (Figure 1.3). Fluidized bed reactors have high heat and mass transfer between feedstock particles and fluidization agents, ensuring constant temperature throughout the gasifier. The operating temperature ranges between 700 to 900 °C (Molino et al., 2018). The major disadvantage of this gasifier is its potential for the slagging of the bed material due to the ash content of the biomass (McKendry, 2002c). Fluidized bed gasifiers are more suitable for large-scale power generation because they can be easily scaled up (Sikarwar et al., 2017).

Entrained Flow Gasifier

Entrained flow gasifiers are mainly used for coal combustion processes. In these gasifiers, the feedstock is fed from the top along with the gasification agent (i.e., O_2) using a burner (Siedlecki et al., 2011). The operating temperature is very high (1200-1500 °C). The high temperature helps in the thermal conversion of tar and methane (Bridgwater, 1995). A biomass pre-treatment is required in this gasification type because of the required feedstock particle size, i.e., fine powder of size 0.1–1 mm (Molino et al., 2016; Siedlecki et al., 2011). This is the major drawback of entrained flow gasifier limiting its use as a chief or solo gasifier.

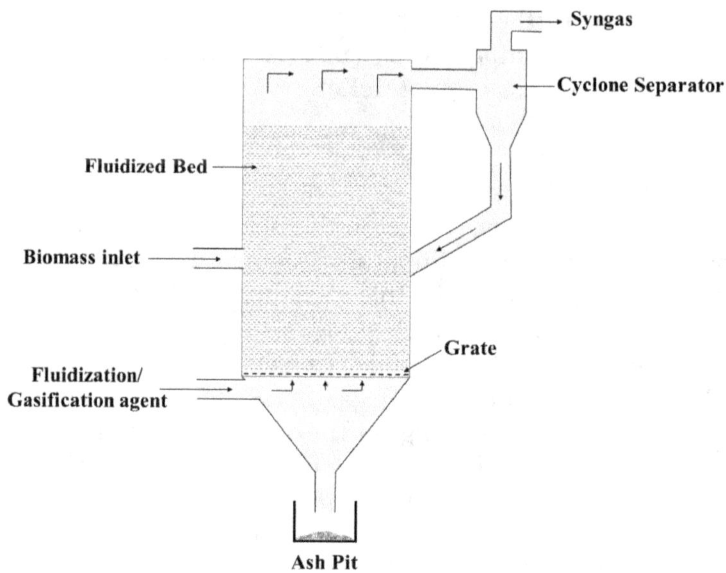

Figure 1.3 Schematic of a fluidized bed gasifier.

Various gasification technologies have their own advantages and limitations. A comparative list of advantages and disadvantages or limitations of all the aforementioned gasification processes is represented in Table 1.1. Amongst the various gasification processes, the downdraft type gasifier is more suitable for comparatively clean biosyngas production.

Table 1.1 Advantages and disadvantages of different types of gasifiers

Type of Gasifier	Advantages	Disadvantages
Updraft	Easy design High quantity of charcoal flame-out. Fuel to gas transformation efficiency is more. High moisture content fuel is accepted. Various size fuels are accepted.	High amount of pyrolysis products and tar. Wide range of gas cleaning required if used for power application.
Downdraft	Less quantity of tar in gas. It is good for the utilization of gas in engines. Fewer particles are available in the gas, at low load conditions.	Scales forming partially More tar produced at low temperature ranges. Higher amounts of ash and dust. Fuel supplies are very difficult.
Cross-draft	It is applicable for small scale processes. At higher temperatures, gas cleaning is not effectively required.	Large quantity of tar produced.
Fluidized bed	Construction is compact. Uniform temperature profile Accepts different fuel size particles. Does not lead to clinker formation at ash melting point of biomass.	Fine dust particles contained by the gas stream. Fuel bed holds less quantity of biomass. So, this system is more complex. Variety of biomass can be applicable, but fuel flexibility is possible for biomass size of 0.1 cm to 1 cm.
Entrained flow	It is applicable for large systems. Very little biomass residence time.	Initial cost is high. Fuel requirements are very strict.

1.3.3 Biosyngas Cleaning and Purification

Biomass gasification is one of the most promising technology for renewable energy generation. Regardless of the significant benefits of biomass gasification and varied applications of syngas, it is yet to attain commercial acceptability. The major bottleneck of this technology is the cleaning of the produced syngas that consists of impurities such as tars, sulphur, particulate matter and ammonia, depending on the process and fuel used (Huang et al., 2011; Thapa et al., 2017).

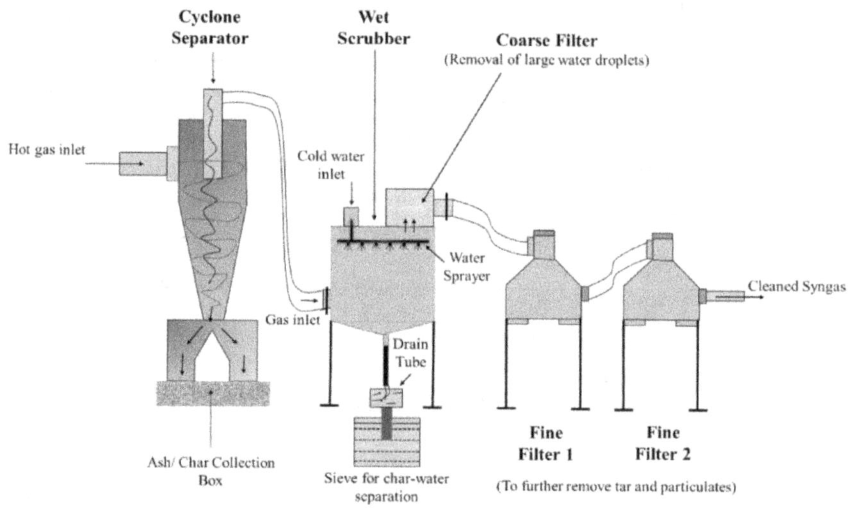

Figure 1.4 Schematic of the syngas cleaning system.

A typical cleaning system of syngas consists of a cyclone separator for the removal of heavy particles, a wet scrubber to remove most of the dust particles and some tar, a course filter for removal of large water droplets, and a few fine filters for further cleaning and purification depending on the quality of syngas produced (Figure 1.4). In the cyclone separator, dust and heavy particles are separated from the syngas by cyclone centrifugal force based on their masses. The gas then enters the wet scrubber from the bottom of the container, and water is sprayed on the incoming gas from the top. Spraying of water forms fine droplets that encapsulate the particulate matter present in the gas, and these droplets are subsequently removed. The temperature of syngas also falls because of the heat transfer between the gas and water. Finally, the clean gas exits the scrubber from the top. Contaminants can be more effectively removed by using different liquids for scrubbing. The course filter later eliminates any traces of moisture from the syngas. The fine filters help in the removal of the remaining tar and water from the gas.

Among the various impurities, tars are the most notorious. They condense and subsequently polymerize on downstream equipment such as compressors and IC engine surface as the gas is cooled down, resulting in system blockage and operational disruption (Zeng et al., 2019). Tar decomposition is attained using either a wet or dry process. The wet process mainly includes syngas scrubbing, as discussed above, using a liquid to absorb the tar products. The dry process includes various methods such as chemical processes, physical filtering and biological methods. The chemical methods include catalytic hydrocracking and thermal cracking of tar compounds contained in the raw syngas. Char is also considered one of the most promising catalysts for tar removal. The chemical process for tar removal has high efficiency, however, it is an energy and cost-intensive process (Babiński and Łabojko, 2014; Zeng et al., 2019). The physical method uses many filters such as biomass filter, sand bed filter, fabric filter, etc., for removal of tar by employing physical compaction, gravitational and electrostatic force, diffusion, etc. Biomass such as wood shavings, dry coconut coir, rice husk, etc., can be used for filtering syngas, with wet scrubber and heat exchanger, to make it acceptable for usage in IC engines. Also, using coarse and fine sand as filter gives good cleaning efficiency. The physical method for tar removal consumes less energy than chemical methods, however, it is less efficient and produces secondary pollutants (Pathak et al., 2007; Thapa et al., 2017; Kumar and Mayilsamy, 2012). The biological method employs active biological enzymes present in the microorganisms to remove tar components through aerobic and anaerobic fermentation processes. This method does not yet have industrial applications because of tar complexity, difficulty in screening high-efficiency bacteria, long treatment time and low removing efficiency (Huang et al., 2011).

1.4 Utilization of Bioenergy

Biomass contains stored energy, making bioenergy a principal source of energy in the field of renewable resources. It contributes roughly one-tenth of the world's total primary energy supply. Moreover, its contribution to the gross global energy demand across all sectors (excluding traditional use of biomass) is five times higher than solar PV and wind combined (IEA, 2020). Bioenergy is widely being used as a fuel for cooking and heating applications worldwide. Also, its use as electricity and transportation fuels is increasing in many developed countries.

Global production of biofuels in 2018 was 154 billion liters, and it is anticipated to increase by 25% during 2019-24, reaching 190 billion liters. In 2018, around 3.7% of transport fuel demand was met by renewable energy (i.e., 4 EJ), out of which biofuel contributed 93%, the remainder being the renewable electricity. As the fossil fuel for transportation is expected to contribute 3% of the total transport fuel demand by 2024, the renewable energy share is expected to increase marginally to 4.6% (i.e., 5.1 EJ), and the biofuel share in it will slightly decrease to 90% (IEA, 2019). Biofuel consumption is expected to increase by 24% (i.e., from 3.7 to 4.6 EJ) in the forecast period (2019-24), while electricity in transport by 70% (i.e., from 0.3 to 0.5 EJ), due to the increased use of electrified rail and electric vehicles (Figure 1.5). Different forms of bioenergy, depending on the type of plant and industrial consumers, can be used for electricity generation. Figure 1.6 shows electricity generation using various bioenergy sources from 1990 to 2018.

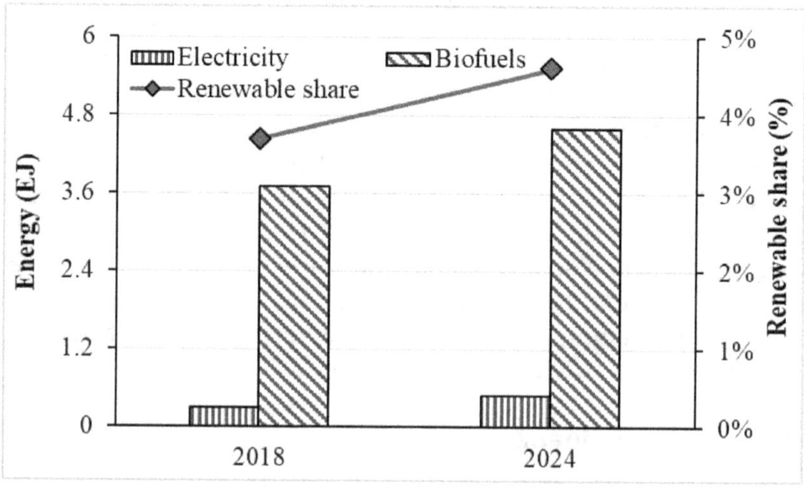

Figure 1.5 Biofuel consumption in the transportation sector in 2018 and 2024 (graph generated using data from IEA, 2019).

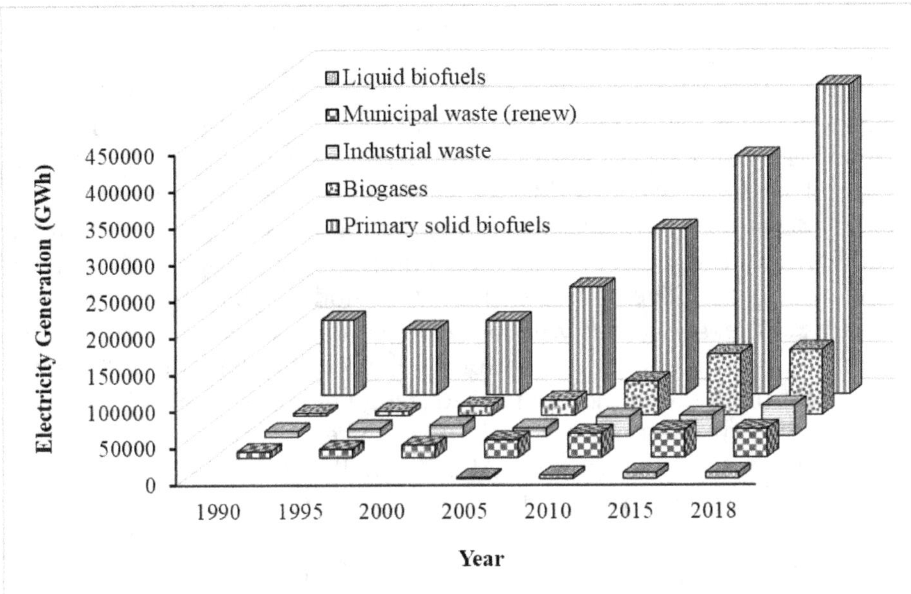

Figure 1.6 Electricity generation from biofuels and waste 1990-2018 (graph generated using data from IEA Renewables Information, 2020).

1.4.1 Application of Biosyngas

Biosyngas is the main product of gasification of biomass, residues and waste. It is produced using biomass and is chemically identical to synthesis gas, hence, termed as biosyngas. Syngas can be used for steam generation in the boiler or treated syngas can be used to run a gas turbine and IC engine.

Biosyngas in Internal Combustion Engines

In the direct use of syngas from a gasifier in an IC engine, the quantity and quality of the gas required are being compromised. So, to maintain the quality, syngas is compressed and stored in gas cylinders for later use in gas turbines and electricity generation. Only a few research studies have focused on compressing and storing syngas, so there is enough room for the researchers to study and produce compressed syngas of significant quality that can be a potential fuel, particularly in small-scale power generation.

The producer gas is a potential and efficient alternative fuel for IC engines. It has a high compression ratio and low rotational speeds (Uppalapati Babu and Kumararaja, 2019). It can be easily compressed to a pressure higher than atmospheric pressure, as it has a poor ignition or a delay in ignition characteristics. So, an appropriate procedure is required to start the ignition of producer gas, as, at the end of the compression stroke, the temperature in the engine is lower than the producer gas's auto-ignition temperature. Therefore, operating the IC engine on dual fuel mode is more profitable in terms of fuel savings as well as a quieter and smoother operation. In a dual-fuel engine, a combination of diesel pilot injection and syngas fumigation in the intake air is used (Singh et al., 2007; Pradhan et al., 2015; Boehman et al., 2008). The producer gas is used as primary fuel and is injected with air. Also, a small quantity of diesel fuel is used for the ignition of producer gas.

Modelling the ignition process of a dual-fuel engine operating with producer gas presents some peculiarity that required the understanding of the heat discharge rate within abnormal combustion. In this dual-fuel combustion, the producer gas is the primary fuel and is introduced along with air. It provides promising energy input to the heat engine. A small quantity of diesel fuel is required to start the ignition of producer gas. The propagation of the combustion zone through the air-producer gas mixtures depends on the mixture's self-ignition temperature (Sombatwong et al., 2013). The schematic diagram of the application of biosyngas in IC engine is shown in Figure 1.7.

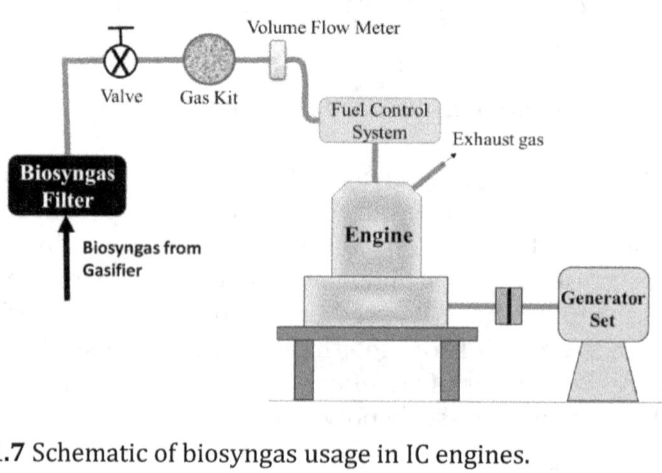

Figure 1.7 Schematic of biosyngas usage in IC engines.

The application of producer gas in IC engine can result in a reduction in petrol and diesel consumption. When producer gas is used as sole fuel in an IC engine, the engine power de-rating is about 15-30%, and emissions to air are reduced (Uppalapati Babu and Kumararaja, 2019).

Biosyngas for Generation of Second-generation Fuels

Second-generation fuels are obtained from various thermochemical processes using biosyngas as raw material. These fuels include both liquid fuels such as ethanol, methanol, Fischer-Tropsch diesel and dimethylether (DME), and gaseous fuels such as synthetic natural gas (SNG) and hydrogen (Demirbas, 2011; Demirbas, 2007; Güell et al., 2013). Various thermochemical processes to achieve different fuels such as H_2, diesel or gasoline, ethanol, methanol/DME and SNG are presented in Figure 1.8 (Spath and Dayton, 2003).

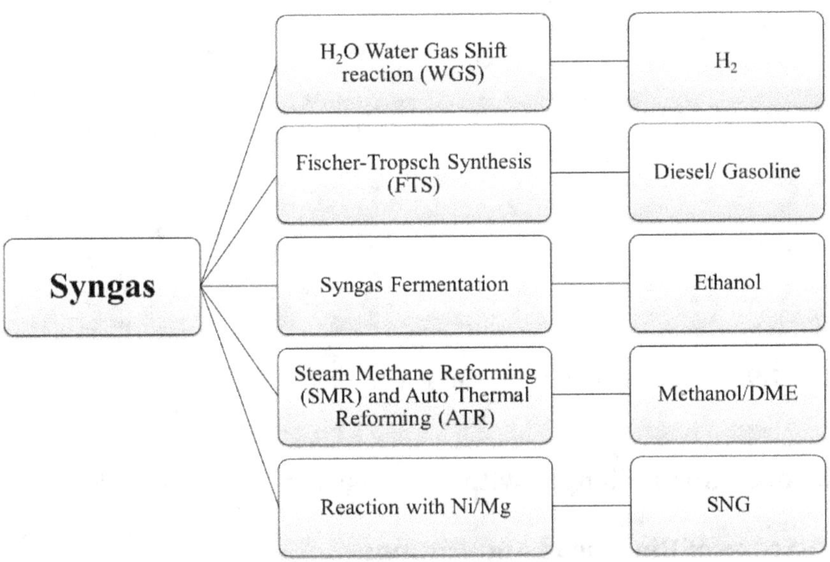

Figure 1.8 Syngas conversion routes to second-generation biofuels.

Miscellaneous Uses of Syngas

Biosyngas can be used for thermal applications in industries or as cooking fuel in residential complexes of hotels. The gas may be pumped at high pressure and either stored in the gas cylinders or pumped to the residential area and/or industries. The use of biosyngas as cooking fuel is shown in Figure 1.9. The biosyngas can be burnt directly for heat generation in specific bakery or drying applications (Panwar et al., 2011).

Apart from the conversion of biomass to biosyngas, the byproducts of biomass gasification can also be used as lubricants, industrial chemicals and many other products currently derived from petroleum or natural gas. More research is required in this regard for exploring further application of byproducts derived during production of biosyngas. An integrated approach is highly appreciated and preferred for sustainable power generation.

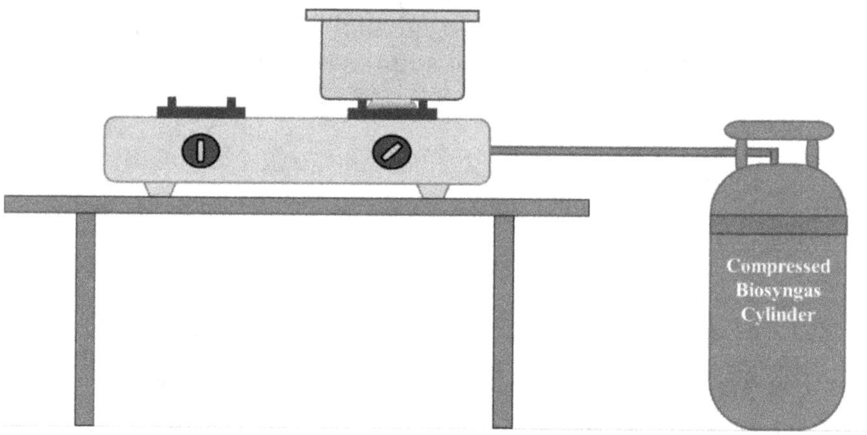

Figure 1.9 Schematic of biosyngas usage in cooking.

1.5 Scope and Challenges with Biosyngas as Sustainable Fuel

1.5.1 Scope of Biosyngas Applications

Biosyngas is produced from biomass and is one of the carbon neutral fuels. Biomass grows in presence of solar energy and CO_2 of the environment. After combustion of biomass derived fuels, the CO_2 is again released into the atmosphere. Hence, biosyngas is one of the sustainable fuels for power generation. Earlier, biomass was used as fuel for cooking application, however, with the development in fossil fuel-based products, i.e., kerosene, liquefied petroleum gas (LPG), piped natural gas (PNG), etc., the use of biomass has decreased. Now, looking at the adverse effects of fossil fuels after combustion, the research community is being compelled for search into alternative sustainable fuels. Therefore, biosyngas, as clean and green fuel, has wide scope for its application in household cooking, power generation and thermal use in industries.

1.5.2 Biomass Availability and Processing Issues

With growing food demand, biomass use in industries for energy generation will inevitably lead to an even fiercer attack on nature. Bioenergy generation will eventually result in the industrialization of forestry and agriculture, resulting in even more deterioration of biodiversity. It would also compete with the food sector, either by using food commodities as an energy source or by using the cultivable land for growing energy crops (Chakravorty et al., 2009). Moreover, biomass collection and transportation for large-scale utilization is cost- and energy-intensive task. As biomass is a widely dispersed and low-density feedstock, careful planning and coordination is required to optimize its movement (Lautala et al., 2015).

A large extent of biomass used for biofuel production is either agricultural waste or is a byproduct of existing agricultural processing activities with very low or no current economic value. Also, biomass dedicated crops are grown on non-agricultural land that otherwise would not produce food. Tapping such biomass resources in the future would break the competition between food and fuel.

1.5.3 Socio-environmental Impact of Biomass Gasification

During the gasification process, many toxic gases are produced, such as CO, SOx, NOx and volatile organics, which can cause serious health hazards (Miguel et al., 2012; Kampa and Castanas, 2008). Hence, a proper leakage check of the entire gasification system and efficient gas cleaning system is essential. Besides health hazards, gasification can also cause fire and explosion because of its operation at very high temperatures and pressure (Luo et al., 2018). The emission of such gases and particulate matter can be controlled by proper leak checking of the unit and using an efficient cleaning system with a conditioning unit. Also, avoiding insufficient combustion and gasification can prevent any such emissions. Fire and explosion are not easily created and can utmost lead to low-quality syngas due to air leakage inside the gasifier.

However, there are downsides too for bioenergy technology. Biofuel production consists of growing energy crops which leads to the clearing of forests. This, in turn, decreases the absorption of carbon dioxide (CO_2) by forests. Also, a lot of energy is consumed in the production processes of biofuel. Greenhouse gases (GHGs) are also generated by the cultivation of energy crops. Therefore, it is clear that the production of biofuel results in GHG emissions, to a greater or lesser extent. On the other hand, CO_2 released during bioenergy production is offset by new plants that consume this CO_2 during growth. These new plants in turn serve as the raw biomass for bio-energy production. Hence, there is a trade-off between the CO_2 reduction when biofuels are used instead of fossil fuels and GHGs that are generated during the production of biofuels. Thus, proper sourcing and processing of biomass for energy are required to make bio-energy a sustainable source of renewable energy.

1.5.4 Challenges for the Use of Producer Gas in IC Engines

The producer gas has a low heating value and contains impurities like tar and dust, which limit its use as fuel in IC engines (Kaupp and Goss, 1981). It contains around 62% of non-combustible components, which do not contribute to its heat value, leading to a low calorific value with low energy density than diesel and natural gas (Ram et al., 2019; Raman and Ram, 2013). Using air-steam mixture as gasification agent and hydrogen enrichment of producer gas can reduce 10% of the non-combustible components. It also lowers the tar content by 5.5%. Hence, using this producer gas in IC engines reduces the specific fuel consumption (SFC) to 20.8% than the SFC of power generation using producer gas from air gasification (Ram et al. 2020).

The improvement in calorific value and reduction in gas storage volume are the challenges. The research work is going on for production and utilization of biosyngas as a sustainable fuel for thermal application and power generation. With time, it may capture space as transportation fuel as well, where production of bio-hydrogen is a promising option with the integrated use of other renewable energies.

Acknowledgement

The authors are grateful to the Scientific and Engineering Research Board, Department of Science and Technology, New Delhi, India, for providing financial support to carry out this work under grant number EEQ/2019/000339.

References

Akia, M., Yazdani, F., Motaee, E., Han, D., Arandiyan, H., 2014. A review on conversion of biomass to biofuel by nanocatalysts. Biofuel Research Journal, Vol. 1(1), pp. 16-25. DOI No: 10.18331/BRJ2015.1.1.5.

Alakangas, Eija, VTT. New European Pellet Standard- EN 14961-1. http://www.infobio.ru/sites/default/files/Alakangas Pellet standard EN14961-1.pdf (Accessed on 22-07-2021).

Babiński, P. and Łabojko, G. (2014). Oxidative thermal cracking of tar compounds in raw syngas from fluidized bed gasification of solid fuels. 6th International Freiberg Conference on IGCC & XtL Technologies, Coal Conversion and Syngas. https://tu-freiberg.de/sites/default/files/media/professur-fuer-energieverfahrenstechnik-und-thermische-rueckstandsbehandlung-16460/publikationen/2014-14-2.pdf (Accessed on 02-08-2021).

Bankhile, K., Basrur, C., Churi, A., Kawediya, Y. (2016). Design And Fabrication of Downdraft Gasifier Applied to I.C. Engine. IOSR Journal of Mechanical and Civil Engineering (IOSR-JMCE) e-ISSN: 2278-1684, p-ISSN: 2320-334X, Vol. 13, Issue 2 Ver. II (Mar. - Apr. 2016), PP 26-28. DOI No.: 10.9790/1684-1302022628.

Barreiro, D. L., Prins, W., Ronsse, F., Brilman, W. (2013). Hydrothermal liquefaction (HTL) of microalgae for biofuel production: state of the art review and future prospects. Biomass and Bioenergy, Vol. 53, pp. 113-127. DOI No.: https://doi.org/10.1016/j.biombioe.2012.12.029.

Basu, P. (2006). Combustion and Gasification in Fluidized Beds. CRC Press: Boca Raton, FL, USA.

Beenackers, A. A. C. M. (1999). "Biomass gasification in moving beds, a review of European technologies". Renewable Energy, Elsevier, Vol. 16(1), pp. 1180-1186. DOI No.: 10.1016/S0960-1481(98)00469-8.

Belgiorno, V., Feo, G. D., Rocca, C. D., Napoli, R. M. A. (2003). Energy from gasification of solid wastes. Waste Management, Vol. 23(1), pp. 1–15. DOI No.: 10.1016/S0956-053X(02)00149-6.

Bioenergy value chain 7: aquatic biomass. European Biofuels Technology Platform. https://etipbioenergy.eu/images/EIBI-7-aquatic%20biomass.pdf (Accessed on 22-07-2021).

Boehman, A. L. and Corre, O. L. (2008). Combustion of syngas in Internal Combustion Engines. Combustion Science and Technology, Vol. 180(6), pp. 1193– 1206. DOI No.: 10.1080/00102200801963417.

Bridgwater, A. V. (1995). The technical and economic feasibility of biomass gasification for power generation. Fuel, Vol. 74, Issue 5, pp. 631-653. DOI No.: 10.1016/0016-2361(95)00001-L.

Bridgwater, A. V. (2012). Review of fast pyrolysis of biomass and product upgrading. Biomass and Bioenergy, Vol. 38, pp. 68-94. DOI No.: https://doi.org/10.1016/j.biombioe.2011.01.048.

Bridgwater, A. V., Hofbauer, H., Loo, S. V. (2009). Thermal Biomass Conversion. Wheelers Books: CPL Press.

Chakravorty, U., Hubert, M. H., Nøstbakken, L. (2009). Fuel versus food. Annual Review of Resource Economics, Vol. 1, pp. 645-663. DOI No.: https://doi.org/10.1146/annurev.resource.050708.144200.

Chawdhury, M. A., and Mahkamov, K. (2010). Development of a Small Downdraft Biomass Gasifier for Developing Countries. Journal of Scientific Research, Vol. 3(1), pp. 51-64. DOI No.: https://doi.org/10.3329/jsr.v3i1.5613.

Chen, H., and Wang, L. (2016). Technologies for Biochemical Conversion of Biomass. Academic Press.

Demirbas, A. (2007). Progress and recent trends in biofuels. Progress in Energy and Combustion Science, Vol. 33, Issue 1, pp. 1-18. DOI No.: https://doi.org/10.1016/j.pecs.2006.06.001.

Demirbas, A. (2011). Competitive liquid biofuels from biomass. Applied Energy, Vol. 88, pp.17–28. DOI No.: 10.1016/j.apenergy.2010.07.016.

Demirbas, A. and Arin, G. (2002). An overview of biomass pyrolysis. Energy Sources, Vol. 24(5), pp. 471–482. DOI No.: 10.1080/00908310252889979.

Garba, A. (2020). Biomass Conversion Technologies for Bioenergy Generation: An Introduction [Online First]. IntechOpen, DOI No: 10.5772/intechopen.93669. https://www.intechopen.com/online-first/73832#B28.

Global Bioenergy Statistics, 2019. World Bioenergy Association. http://www.worldbioenergy.org/uploads/191129%20WBA%20GBS%20 2019_HQ.pdf (accessed on 06-07-2021).

Global biomass potential towards 2035 (2016). WBA fact sheet. http://www.worldbioenergy.org/uploads/Factsheet_Biomass%20potential.pdf (accessed on 20-07-2021).

Güell, B. M., Sandquist, J., Sørum, L. (2013). Gasification of Biomass to Second Generation Biofuels: A Review. Journal of Energy Resources Technology- transactions of the ASME, Vol. 135 (1), pp. 14001. DOI No.: https://doi.org/10.1115/1.4007660.

Horan, N., Yaser, A., Wid, N. (Eds.) (2018). Anaerobic Digestion Processes. Dalena, F., Senatore, A., Tursi, A., Basile, A., 2017. Bioenergy production from second- and third-generation feedstocks, in: Dalena, F., Basile, A., Rossi, C. (Eds.). Bioenergy Systems for the Future: Prospects for Biofuels and Biohydrogen. Elsevier Publishing, London, pp. 559-599. Green Energy and Technology. Springer, Singapore, pp 1-7. ISSN: 1865-3529.

Hornung, A. (2014). Transformation of Biomass: Theory to Practice. John Wiley & Sons.

Hornung, A., Apfelbacher, A., Sagi, S. (2011). Intermediate pyrolysis: A sustainable biomass-to-energy concept- biothermal valorisation of biomass (BtVB) process. Journal of Scientific and Industrial Research, Vol. 70, pp. 664-667.

Huang, J., Schmidt, K. G., Bian, Z. (2011). Removal and Conversion of Tar in Syngas from Woody Biomass Gasification for Power Utilization using Catalytic Hydrocracking. Energies, Vol. 4, pp. 1163-1177. DOI No.: 10.3390/en4081163.

IEA (2019). Renewables 2019. International Energy Agency, Paris. https://www.iea.org/reports/renewables-2019 (Accessed on 23-07-2021).

IEA (2020). Bioenergy. International Energy Agency, Paris. https://www.iea.org/fuels-and-technologies/bioenergy (Accessed on 23-07-2021).

IEA Renewables Information (2020). International Energy Agency. https://www.iea.org/fuels-and-technologies/renewables (Accessed on 23-07-2021).

Jain, A. K. (1999). Biomass gasification under oxygen medium. Journal of Agricultural Engineering, Vol. 36 (3), pp. 27–31.

Jindal, S., Salvi, B. L. (2012). Sustainability aspects and optimization of linseed biodiesel blends for compression ignition engine. Journal of Renewable Sustainable Energy 4: 043111. http://dx.doi.org/10.1063/1.4737922.

Kampa, M. and Castanas, E. (2008). Human health effects of air pollution. Environmental Pollution, Vol. 151(2), pp. 362-367. DOI No.: 10.1016/j.envpol.2007.06.012.

Kaupp, A. and Goss, J. R. (1981). State of the art report for small scale (to 50 kW) gas producer engine systems. Canada: Department of agricultural engineering, pp. 278.

Kumar, R. R. and Mayilsamy, K. (2012). A novel compact Bio-filter system for a downdraft gasifier: An experimental study. AASRI Procedia, Vol. 3, pp. 700-706. DOI No.: 10.1016/j.aasri.2012.11.111.

Kunii, D. and Levenspiel, O. (1991). Fluidization Engineering. Butterworth-Heinemann: Oxford, UK.

Lauri, P., Havlík, P., Kindermann, G. E., Forsell, N., Böttcher, H., Obersteiner, M. (2014). Woody biomass energy potential in 2050. Energy Policy, Vol. 66, pp. 19-31. DOI No.: 10.1016/j.enpol.2013.11.033.

Lautala, P. T., Hilliard, M. R., Webb, E., Busch, I., Hess, J. R., Roni, M. S., Hilbert, J., Handler, R. M., Bittencourt, R.,Valente, A., Laitinen, T. (2015). Opportunities and Challenges in the Design and Analysis of Biomass Supply Chains. Environmental Management, Vol. 56(6), pp. 1397–1415. DOI No.: 10.1007/s00267-015-0565-2.

Li, Y., Chen, Y., Wu, J. (2019). Enhancement of methane production in anaerobic digestion process: A review. Applied Energy, Elsevier, Vol. 240(C), pp. 120-137. DOI No.: 10.1016/j.apenergy.2019.01.243.

Luo, X., Wu, T., Shi, K., Song, M., Rao, Y. (2018). Biomass Gasification: An Overview of Technological Barriers and Socio-Environmental Impact. Gasification for Low-grade Feedstock, Yongseung Yun, IntechOpen. DOI No.: 10.5772/intechopen.74191. Available from: https://www.intechopen.com/chapters/59423#B24.

Mahalaxmi, S. and Williford, C. (2014). Biochemical conversion of biomass to fuels, in: Chen, W., Suzuki, T., Lackner, M. (Eds.), Handbook of climate change mitigation and adaptation. Springer, New York, pp. 1-28.

Mai, T. P. and Nguyen, D. Q. (2020). Gasification of Biomass [Online First]. IntechOpen, DOI No.: 10.5772/intechopen.93954. Available from: https://www.intechopen.com/online-first/73450.

McKendry, P. (2002). Energy production from biomass (part 1): Overview of biomass. Bioresource Technology, Vol. 83(1), pp. 37–46. DOI No.: 10.1016/S0960-8524(01)00118-3.

McKendry, P. (2002). Energy production from biomass (part 2): Conversion technologies. Bioresource Technology, Vol. 83(1), pp. 47–54. DOI No.: 10.1016/S0960-8524(01).

McKendry, P. (2002). Energy production from biomass (part 3): Gasification technologies. Bioresource Technology, Vol. 83(1), pp. 55-63. DOI No.: 10.1016/s0960-8524(01)00120-1.

Miguel, G. S., Domínguez, M. P., Hernández, M., Sanz-Pérez, F. (2012). Characterization and potential applications of solid particles produced at a biomass gasification plant. Biomass and Bioenergy, Vol. 47, pp. 134-144. DOI No.: 10.1016/j.biombioe.2012.09.049.

Mohammed, N. I., Kabbashi, N., Alade, A. (2018). Significance of Agricultural Residues in Sustainable Biofuel Development, Agricultural Waste and Residues, Anna Aladjadjiyan, IntechOpen, DOI No.: 10.5772/intechopen.78374. Available from: https://www.intechopen.com/chapters/62260.

Molino, A., Chianese, S., Musmarra, D. (2016). Biomass gasification technology: the state of the art overview. Journal of Energy Chemistry, Vol. 25, Issue 1, pp. 10-25. DOI No.: 10.1016/j.jechem.2015.11.005.

Molino, A., Larocca, V., Chianese, S., Musmarra, D. (2018). Biofuels Production by Biomass Gasification: A Review. Energies, Vol. 11, pp. 811. DOI No.: 10.3390/en11040811.

Mukhtar, S. and Capareda, S. (2017). Manure to Energy: Understanding Processes, Principles and Jargon. The Texas A&M University System. https://tammi.tamu.edu/2017/07/17/manure-energy-understanding-processes-principles-jargon/ (Accessed on 22-07-2021).

Panwar NL, Salvi BL, Reddy VS. Performance evaluation of producer gas burner for industrial application. Biomass and Bioenergy 2011; 35(3):

1373–1377. http://dx.doi.org/10.1016/j.biombioe.2010.12.046.

Pathak, B. S., Kapatel, D. V., Bhoi, P. R., Sharma, A. M., Vyas, D. K. (2007). Design and Development of Sand Bed Filter for Upgrading Producer Gas to IC Engine Quality Fuel. International Energy Journal, Vol. 8, pp. 15-20.

Pradhan, A., Baredar, P., Kumar, A. (2015). Syngas as an Alternative Fuel Used in Internal Combustion Engines: A Review. Journal of Pure and Applied Science and Technology, Vol. 5(2), pp. 51-66.

Prasad, L., Salvi, B. L., Kumar, V. (2015). Thermal degradation and gasification characteristics of Tung Shells as an open top downdraft wood gasifier feedstock. Clean Technologies and Environmental Policy 17(6), pp. 1699-1706. http://dx.doi.org/10.1007/s10098-014-0891-8.

Rajvanshi, A. K. (1986). Biomass Gasification, in: Alternative Energy in Agriculture, Y. Goswami (ed.), Vol. II (CRC Press, U.S.A., 1986).

Ram, N. K., Singh, N. R., Raman, P., Kumar, A., Kaushal, P. (2019). A detailed experimental analysis of air-steam gasification in a dual fired downdraft biomass gasifier enabling hydrogen enrichment in the producer gas. Energy, 187:115937. DOI No.: 10.1016/j.energy.2019.115937.

Ram, N. K., Singh, N. R., Raman, P., Kumar, A., Kaushal, P. (2020). Experimental study on performance analysis of an internal combustion engine operated on hydrogen-enriched producer gas from the air-steam gasification. Energy, Vol. 205, pp. 118029. DOI No.: https://doi.org/10.1016/j.energy.2020.118029.

Raman, P. and Ram, N. K. (2013). Performance analysis of an internal combustion engine operated on producer gas, in comparison with the performance of the natural gas and diesel engines. Energy 2013, 63:317e33. DOI No.: https://doi.org/10.1016/j.energy.2013.10.033.

Rasmussen, M. B., Jensen, P. D., Bruhn, A., Mouritsen, L. T., Olesen, B., Markager, S., Thomsen, A. B., Sander, B. (2009). Aquatic biomass for biofuels. IOP Conference Series: Earth and Environmental Science, Vol. 6, 182010. DOI No.: 10.1088/1755-1307/6/18/182010.

Reed, T.B. and Das, A. (1988). Handbook of Biomass Downdraft Gasifier Engine Systems. Solar Technical Information Program. Solar Energy Research Institute. SERIISP-271-3022 DE88001135 March 1988 UC Category.' 245.

Salvi, B.L., Soni, T., Jindal, S., Panwar, N.L. (2021). Design improvement and experimental study on shell and tube condenser for bio-oil recovery from fast pyrolysis of wheat straw biomass. SN Applied Sciences 3: 173, https://doi.org/10.1007/s42452-021-04165-8, ISSN: 2523-3971.

Sharma, S., Meena, R., Sharma, A., Goyal, P. K. (2014). Biomass Conversion Technologies for Renewable Energy and Fuels: A Review Note. IOSR Journal of Mechanical and Civil Engineering (IOSR-JMCE), e-ISSN: 2278-1684, p-ISSN: 2320-334X, Volume 11, Issue 2 Ver. I (Mar- Apr. 2014), PP 01-00. DOI No.: 10.9790/1684-11232835.

Sharma, V.K., 2015. Technology development and innovation for production of next-ge neration biofuel from lignocellulosic wastes, in: Sharma, A., Kar, S. (Eds.), Energy sustainability through green energy. Green Energy Technology. Springer, New Delhi, pp. 315-350.

Siedlecki, M., Jong, W. D., Verkooijen, A. H. M. (2011). Fluidized Bed Gasification as a Mature and Reliable Technology for the Production of Bio-Syngas and Applied in the Production of Liquid Transportation Fuels—A Review. Energies, Vol. 4, pp. 389-434. DOI No.: 10.3390/en4030389.

Sikarwar, V. S., Zhao, M., Fennell, P. S., Shah, N., Anthony, E. J. (2017). Progress in biofuel production from gasification. Progress in Energy and Combustion Science, Vol. 61, pp. 189–248. DOI No.: 10.1016/j.pecs.2017.04.001.

Singh, R. N., Singh, S. P., Pathak, B. S. (2007). Performance of renewable fuel-based CI engine. International Commission of Agricultural Engineering, CIGR E-Journal, Vol. 9, pp. 1-10.

Sombatwong, P. et al. (2013). Effect of Pilot Fuel Quantity on the Performance and Emission of a Dual Producer Gas - Diesel Engine. Energy Procedia, Vol. 34, pp. 218 – 227. DOI No.: 10.1016/j.egypro.2013.06.750.

Spath, P. L. and Dayton, D. C. (2003) Preliminary Screening—Technical and Economic Assessment of Synthesis Gas to Fuels and Chemicals with Emphasis on the Potential for Biomass-Derived Syngas. National Renewable Energy Laboratory NREL/TP-510-34929.

Stamatov, V., Honnery, D., Soria, J. (2006). Combustion properties of slow pyrolysis bio-oil produced from indigenous Australian species. Renewable Energy, Vol. 31(13), pp. 2108-2121. DOI No.: 10.1016/j.renene.2005.10.004.

Strezov, V., 2014. Properties of biomass fuels, in: Strezov, V., Evans, T. J. (Eds.). Biomass processing technologies. CRC Press, Boca Raton, pp. 1-32.

Tanger, P., Field, J. L., Jahn, C., Defoort, M., Leach, J. (2013). Biomass for thermochemical conversion: Targets and challenges. Frontiers in Plant Science, Vol. 4, pp. 218. DOI No.: 10.3389/fpls.2013.00218.

Thapa, S., Bhoi, P. R., Kumar, A., Huhnke, R. L. (2017). Effects of Syngas Cooling and Biomass Filter Medium on Tar Removal. Energies, Vol. 10, pp. 349. DOI No.: 10.3390/en10030349.

Tursi, A. (2019). A review on biomass: importance, chemistry, classification, and conversion. Biofuel Research Journal, Vol. 22, pp. 962-979. DOI No.: 10.18331/BRJ2019.6.2.3.

Uppalapati, B. and Kumararaja, L. (2019). A Research on Producer Gas in Internal Combustion Engines. International Journal of Recent Technology and Engineering (IJRTE), Vol. 8, ISSN: 2277-3878, Issue-1S4.

Verma, A. and Mondal, P. (2016). Pyrolysis of sugarcane bagasse in semi batch reactor: Effects of process parameters on product yields and characterization of products. Industrial Crops and Products, Vol. 95, pp. 704-717. DOI No.: 10.1016/j.indcrop.2016.11.039.

World energy needs. (2021). Canada's Oil and Natural Gas Producers. https://www.capp.ca/energy/world-energy-needs/ (accessed on 06-07-2021).

Zamani, A. (2015). Introduction to lignocellulose-based products, in: Karimi, K. (ed.), Lignocellulose-Based Bioproducts. Springer, Cham, pp. 1-36.

Zeng, X., Ueki, Y., Yoshiie, R., Naruse, I., Wang, F., Han, Z., Xu, G. (2019). Recent progress in tar removal by char and the applications: A comprehensive analysis. Carbon Resources Conversion, Vol. 3, pp. 1–18. DOI No.: https://doi.org/10.1016/j.crcon.2019.12.001.

2

Smart Grid and Its Current and Future Applications

Ankan Bhunia

NITTTR, Chandigarh, India

Abstract

Smart grid is a concept which upgrades the conventional electrical power grid by adopting automatic control devices, implementing smart communications technologies, analyzing real time data and applying software-based solution for continuous, efficient and cost-effective power transmission and distribution. In this chapter, the concept of smart grid and its overview are explained broadly, and its advantages and disadvantages as well as current and future applications are presented.

2.1 Introduction

In the conventional method of transmission and distribution of electrical energy, high voltage electricity is transmitted from the electrical power generation plant to the substations. After that, the voltage level is stepped down using step down transformer for distributing the electrical energy to the end-users. In the last few years, there has been a significant use of the renewable energy sources in the generation of electrical energy, thus, requiring a smart electrical distribution system for efficient and uninterrupted electricity supply.

As the society and economy are growing, the energy resources are reducing, and the environment becomes more polluted, therefore, we need to ensure a clean energy supply.

Smart grid is the concept that provides a solution for these problems. The way a smart grid is constructed increases the adaptability of clean energy in the power grid, thus, providing a secure, efficient and economical power supply.

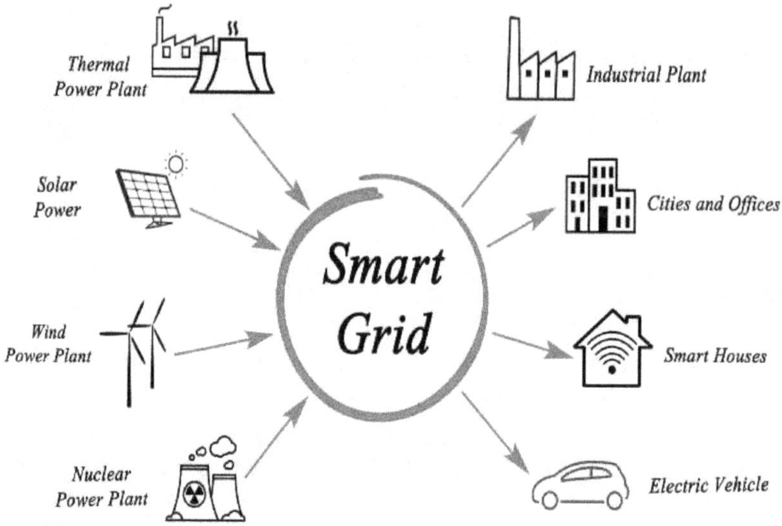

Figure 2.1 Smart grid [1].

In the modern world, due to the renewable energy sources, the consumers can contribute to the energy supply (Figure 2.1). Thus, as per the demand, we need a bidirectional power grid system that can ensure two-way communication between the supplier and consumer. Smart grid can ensure two-way communication and uses computer processing as well as advanced control for automation. The phasor measurement units (PMUs) for ensuring the grid stability, advanced digital meters, PLC, SCADA, etc., are used to obtain the real-time data, accurate measurements and automatic outage reporting. Further, microprocessor-based relay is used to sense and recover from faults. The automatic feeder switches are used to re-route the power line in case any problem arises in the line, so as to ensure continuous power supply [2].

A smart grid is *"a modernized grid that enables bidirectional flows of energy and uses two-way communication and control capabilities that will lead to an array of new functionalities and applications."* - NIST

"The smart grid is the electricity delivery system, from point of generation to point of consumption, integrated with communications and information technology for enhanced grid operations, customer services and environmental benefits."
- U.S. Dept. of Energy

"The smart grid has come to describe a next-generation electrical power system that is typified by the increased use of Communications and Information Technology in the generation, delivery and consumption of electrical energy."
- IEEE [3]

2.2 Need of Smart Grid

In today's world, the use of electricity is vastly spreading in different areas, like electric vehicles (EVs), which would impose a significant load on the network in the coming years. So, the electricity demand is rapidly increasing all over the world. For instance, the data from Central Electricity Authority of the Government of India is shown below (Figure 2.2).

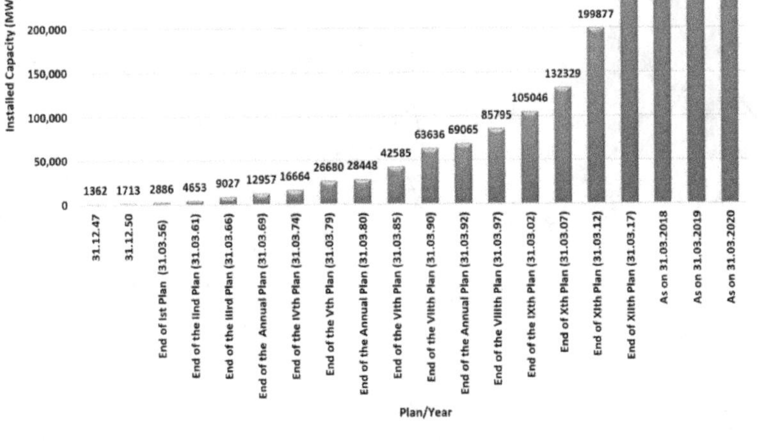

Figure 2.2 Plan wise growth of installed capacity in India [4].

Looking at the data, we can imagine how the electricity demand is increasing every year, and, in the year 2020, the installed capacity of electricity was 370106 MW [4]. Thus, problems like overload, short circuits and others faults in the network are going to become common. For this purpose, we need a well-planned smart grid system that can transfer and distribute electricity efficiently and economically.

The conventional power generation is not environment friendly, so we need a smart grid system to reduce the power loss for the reduction of environmental pollution. As the consumer also becomes a source, so the two-way communication between the consumer and grid is needed, and a smart grid can provide such features. Apart from these, a smart grid will provide superior security, efficiency enhancement and automation.

2.3 Structure of Smart Grid

Due to the advancement in the power sector, the power system is transforming from centralized tropology to decentralized and distributed topology [5].

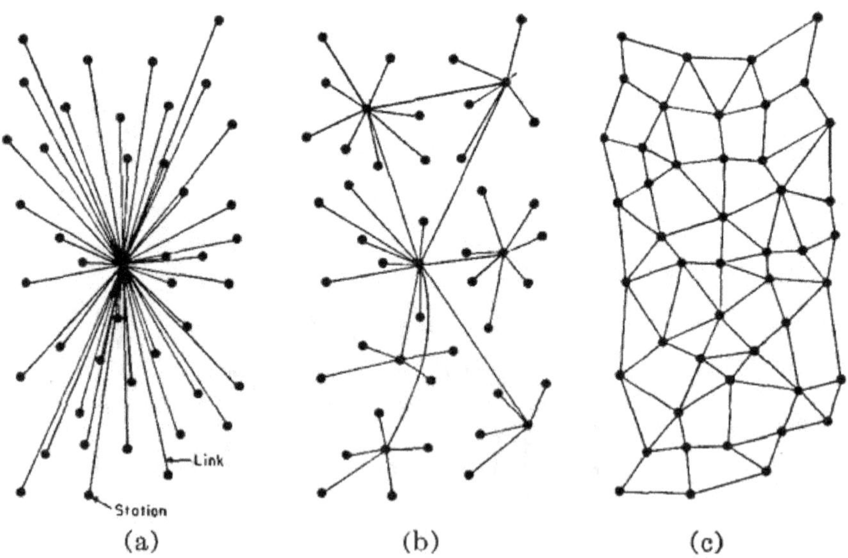

Figure 2.3 Power generation topologies [5]: (a) centralized, (b) decentralized and (c) distributed networks.

With enhanced use of wind turbines, photovoltaic solar energy systems and micro-turbines, various types of batteries for energy storage and various small scale power generation plants have been established. In contrast with the other conventional energy sources like thermal power plants, nuclear power plants, etc., these units are connected in the electric network. Thus, a grid system is needed to ensure a secure and reliable transformation of measurement and control the information between the sources and consumers and vice versa.

New smart grid architectures can incorporate volatility in energy provision and decentralized topological energy generation. The smart grid architecture has two-way communication, computing and microprocessors-based automatic digital electronics measurement systems [6].

Smart grid elements are as follows [6]:

- Power generation
- Transmission system
- Distribution system
- Different renewable energy sources
- Storage system
- Residential, industrial and commercial customers

A common representation of the smart grid architecture is shown below:

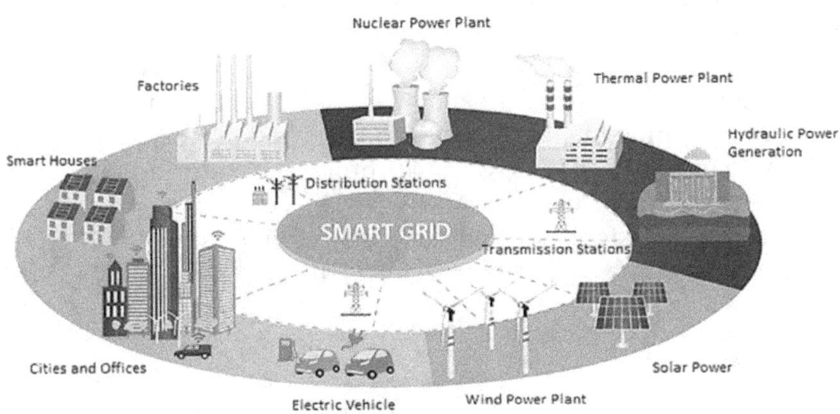

Figure 2.4 Smart grid architecture [6].

2.4 Operation/Management of Smart Grid

The management of smart grid is attained by the continuous monitoring of the network, followed by the optimization of the collected data and taking decision based on the optimization results automatically. Smart grid management is run by implementing the steps mentioned below [6]:

- Smart metering: Monitoring and helps in two-way communication.
- Efficient management of conventional energy sources as well as renewable energy sources.
- Good management of power transmission and distribution between the main and microgrid.
- Smart management for mobile electricity consumers [6] (e.g., EVs).

In a smart grid management system, the real-time monitoring of the network is ensured, and the information is sent to the respective governing body. Based on this, the smart grid is managed efficiently and securely.

2.5 Communication in Smart Grid

Due to the advancement of the renewable energy sources, a consumer can now be a contributor to power network. As a result, the distributed decentralized energy network is growing rapidly, where we need to ensure the two-way communication between the grid and consumer.

For good communication in the smart grid, we need to mention the parameters. By implementing the regulations and protocols, a standard can be achieved. The IPv6 routing protocol for metering substructure is used as a protocol in the smart grid [7]. For secure and continuous data transmission, the smart meter reading data should have data protection of the consumer as well as the governing body [5].

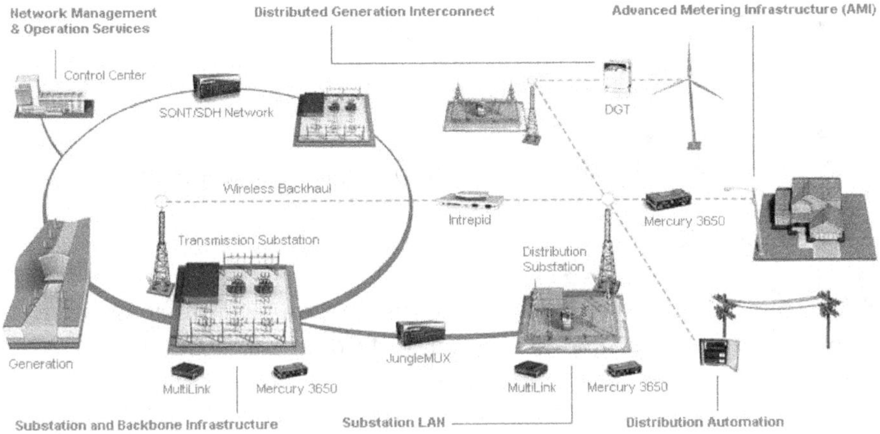

Figure 2.5 Smart grid communication [8].

In case of an issue, the smart grid should have the required software and storage for retrieving the data.

In the smart grid, various types of digital electronics devices are connected with the latest sensors, along with the other security purpose devices to ensure efficient, secure and reliable data transmission between the various connected systems (Figure 2.5).

2.6 Technology Used in Smart Grid

2.6.1 Advanced Metering Infrastructure

Advanced metering infrastructure (AMI) is a system that collects, measures and analyzes energy usage by enabling data to communicate over a two-way communications network connected with the advanced meters ("smart meters") and utility's control systems [9]. Uses of AMI allows the end-users to make informed and real-time choices about their energy usage. The main components of the AMI system are as follows:

- Smart meter, which can communicate in a bidirectional way.
- A secure communications network that can transmit the measured data of the smart meter [9].

AMI has other important applications [9]:

- Remote meter reading
- Remote controlling
- Smart outage detection and management
- Tamper detection
- Distributed generation, monitoring and management

2.6.2 Smart Meter

Smart meter is a very important device in smart grid which helps to truly define the word 'smart' in real-life operation. Compared with the conventional meter, smart meters are more like sensors in energy networks [11]. These measure the power and store the data continuously at some time interval. The data is transferred to the governing body through a secure communication system. The data helps to analyze the behavior of the system, secure continuous power supply and attain fast outage response. If any disturbance occurs in the system, then the smart meter will send notifications to the users and governing body through direct massaging or any specific app.

In short, a smart meter ensures two-way communication between the consumers and governing body. It collects, measure and stores the data securely for informing the system as well as the consumer about an outage. It also performs the billing at the month-end or quarter-end automatically.

2.6.3 Distributed Generation

Due to the increasing use of renewable energy sources, various micro- and mid-range power generation plants are running in various places. Using smart grid technology, these distributed generation plants can contribute to the main energy network where smart grid technology automatically measures, monitors, controls and distributes the power efficiently.

2.6.4 Phasor Measurement Units

PMU, also known as 'synchrophasors', provide the operators a time-stamped snapshot of the power system using a common time source for synchronization [10]. These can calculate real and reactive power, frequency and phase angle 12 times per 60 Hz cycle [10]. Basically, PMU collect the real-time data by measuring or sampling voltage and current at a given location of the network for a certain period which helps the monitoring of the power system more accurately and efficiently.

2.7 Advantage and Disadvantages of Smart Grid

Advantages of a smart grid are given below:

Energy saving: Due to the use of smart energy meters and other less power-consuming devices, a smart grid uses less amount of energy, thus, leading to energy saving.

Self-healing [10]: By using real-time sensors, smart grid can search, detect and react if any disturbance occurs and is able to start, stop or relocate the power flow to ensure continuity.

Secure network: Smart grid uses the latest sensors, measuring devices and other cyber security-related systems to prevent the network from cyber-attack.

Environment friendly: Smart grid has the capability to transfer and distribute power in the most efficient way, so the wastage of energy is less, which ultimately reduces pollution. Apart from that, managing renewable energy distribution is easy in smart grid system. As a result, the use of renewable energy sources has increased. Smart grid maintains an efficient demand-supply ratio between the sources and consumer, so the wastage of energy is reduced, which prevents the environment from more coal burning.

Economical: Smart grid allow a consumer to become a supplier which reduces the overall price of electricity. Using smart grid power generation, companies get a better idea about the demand for power in the market. It encourages companies to establish more efficient equipment and find efficient energy production techniques.

Distributed generation [10]: Smart grid encourages distributed generation by which reduction of pressure on the main grid happens during the peak time. If any disturbance happens in the main grid, the distributed generation plants can continue the power flow. It also reduces the cost of power transmission.

Bidirectional communication: Due to the use of smart electronic devices in a smart grid, it can provide us the bidirectional data and power flow in the network. It introduces the net metering concept with the consumers by which consumers can contribute to the power grid easily.

Carbon emissions reduction: Smart grid reduces the consumption of electricity. As a result, the carbon emission is reduced.

The disadvantages of a smart grid are given below:

Economical investment: Although the operation in a smart grid is economical, but the initial installation cost is significant.

Security and privacy: In smart grid, different types of programmable microprocessor or microcontroller-based electronic devices are used. In addition, for data storage, various cloud services are used. These digital electronic devices can be hacked, thus, the smart grid suffers from the issues such as cyber security and data privacy of the consumer.

Availability of equipment: Most of the electronic devices used in the smart grid are the latest ones. Sometimes a shortage of various devices can be there.

Not consumer friendly: The concept of a smart grid is not consumer friendly. Sometimes, consumers do not completely understand the net metering concept and other protocols.

Highly skilled workers: To operate a smart grid, highly skilled workers are needed. So, a smart grid companies need to provide skill training to the new workers, which adds to the cost.

Complex system: Smart grid is a very complex system, which is hard to understand and operate. Its maintenance is costly and complex, thus, requiring highly skilled workers and complex machinery.

Continuous upgradation: In a smart grid, microprocessor-based electronics devices are used, which are updated almost every year. For security purposes, we need to update the equipment in the system regularly, thus, making the smart grid operation costly.

2.8 Comparison between Conventional Grid and Smart Grid

Sl. No.	Conventional grid	Smart grid
1	A conventional grid is a system through which the power is controlled and managed by centralized energy topology.	A smart grid is a system that controls and manages the power by decentralized and distributed topology.
2	In a conventional grid system, most of the equipment is based on electromechanical technology.	In a smart grid system, most of the equipment is based on microprocessor-based digital technology.
3	In a conventional grid, power loss is more compared with a smart grid.	Power loss in the smart grid is less.
4	Control on energy distribution after it is out from the generation plant is less in a conventional grid.	Control on energy is possible throughout the system in the smart grid.
5	Conventional grid is a unidirectional system, so the communication with the consumer is limited.	Smart grid is a bidirectional system, so the consumers also participate in the main grid power supply.
6	In conventional grid, integration of renewable energy is completive and less efficient.	In smart grid, integration of renewable energy is easy, which encourage the people to use renewable energy. As a result, it reduces pollution.
7	Others: More maintenance. Cannot be controlled from a remote location. For future modification, the hardware is needed to be changed. Less stable during disturbance. Less efficient.	Others: Less maintenance. Can be controlled from a remote location. Future modification can be performed by modifying the program without changing the hardware for a specific application. More stable during disturbance. More efficient.

2.9 Current and Future Applications of Smart Grid

2.9.1 Big Data Analysis in Smart Grid

In today's world, most of the devices are powered by electricity, and the electric power is managed and controlled by the power grid. In the smart grid, various types of sensors are used which generate a large amount of data. Apart from this, other non-electrical data like GIS, weather reports, etc., are gathered by the various social media, PLC, SCADA, etc., called collectively as the big data. The amount of data is significant. For instance, the amount of data was 33 zettabytes (ZB) in the year of 2018 [12], which is equivalent to 33 trillion gigabytes. So, it is very important to analyze the data for better performance of the smart grid.

Big data analysis has various applications in different fields of a smart grid: to extract the important and meaningful information from the generated data, followed by its analysis for the greater good.

Big data analysis is applied in various parts of smart grid. For instance, in generation part, big data analysis helps in generation planning, load forecasting, etc., useful for health monitoring of smart grid equipment and real-time energy management. It also helps in understanding the customer behaviors, market monitoring, stability enhancement, etc.

By applying big data analysis in the smart grid, we can reduce the wastage of energy in various ways which ultimately fulfills the green and clean energy goals.

Here, an example of the use of big data analysis in a smart grid is presented. A home energy report [13] of a customer published by PG&E demonstrates that the big data analysis helps the customer to make decisions about smart energy saving. By looking at the report, the consumer can know if the use of power was more/less in a month. In addition, the big data analysis helps the consumer by suggesting the possible ways for energy saving (Figure 2.6).

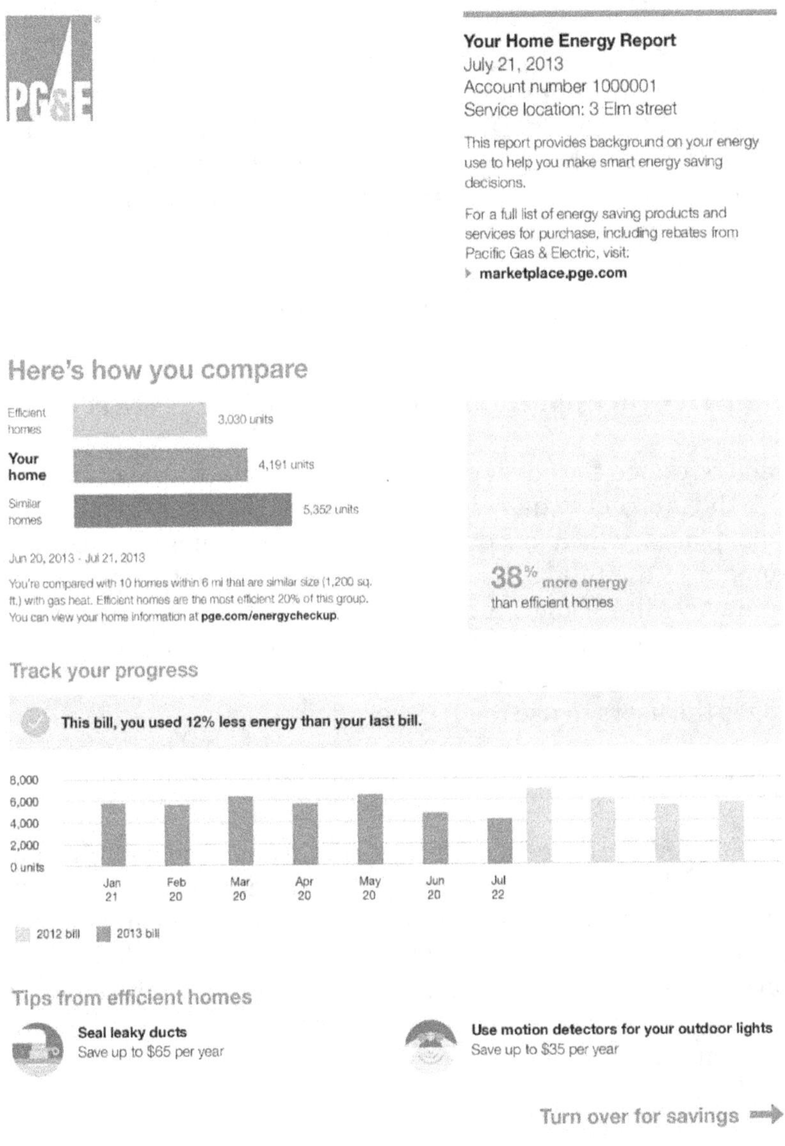

Figure 2.6 A home energy report by PG&E [13].

2.9.2 EV Integration in Smart Grid

As the energy sources of conventional vehicles are limited and ultimately lead to the environmental pollution, the use of electric vehicles has attracted the consumer's interest. In future, a majority of the transportation will be based on EVs. By using the smart meters and other advanced devices in the smart grid, we can use EVs as a source of energy if we implement solar panels or other renewable energy sources on these [14], termed as the plug-in hybrid electric vehicle technology (PHEV). Other than that, the smart grid can make an efficient and reliable power management system for EV charging.

2.9.3 Smart Grid Application on Internet of Things (IOT)

Nowadays, due to the advancement of internet technology, most people use the internet in daily life. 5G network is already established in various places which encourages the advancement of IOT based equipment in the future. By use of IOT, we can control various electronic devices like fridges, air coolers, TV, etc., from a remote location. By using a smart grid, the IOT technology can be implemented more securely. The smart grid can protect the data of the users, which encourages the users to use the IOT based products.

2.10 Conclusion

In this chapter, the concept of the smart grid and some possible future applications have been discussed. Although the initial cost of smart grid establishment is high, however, the smart grid can play a very crucial role for efficient energy management and for achieving clean and green energy goals. The energy management in future will completely transform from conventional to smart and secure by the use of the smart grid.

References

1. http://smartgrid.ieee.org.
2. Ilhami Colak (2016) "Introduction to smart grid". 2016 International Smart Grid Workshop and Certificate Program (ISGWCP), Istanbul, p. 30.
3. http://smartgrid.ieee.org.
4. https://cea.nic.in.

5. Ilhami Colak (2016) "Introduction to smart grid". 2016 International Smart Grid Workshop and Certificate Program (ISGWCP), Istanbul, p. 31.
6. Boban Panajotovic, Milan Jankovic, Borislav Odadzic (2011) "ICT and Smart Grid" .2011 TELSIKS, Serbia, pp. 119-120.
7. D. Wang, Z. Tao, J. Zhang, and A. a Abouzeid, "RPL based routing for advanced metering infrastructure in smart grid," Commun. Work. (ICC), 2010 IEEE Int. Conf., pp. 2–6.
8. V. C. Güngör, D. Sahin, T. Kocak, S. Ergüt, C. Buccella, C. Cecati, and G. P. Hancke, (2011) "Smart grid technologies: Communication technologies and standards," IEEE Trans. Ind. Informatics, vol. 7, no. 4, pp. 530.
9. David J. Leeds, 'The Smart Grid in (2010): Market Segments, Applications and Industrial Players', GTM Research, July 2009.
10. Md. Golam Rahman, M. Fahad Bin Ramim Chowdhury, Md. Abdulla Al Mamun, Md. Rakib Hasan, Sayeed Mahfuz (2013) "Summary of Smart Grid: Benefits and Issues", International Journal of Scientific & Engineering Research Volume 4, Issue 3, pp. 1-4.
11. Gouri R. Barai, Sridhar Krishnan, Bala Venkatesh (2015), "Smart Metering and Functionalities of Smart Meters in Smart Grid - A Review", 2015 IEEE Electrical Power and Energy Conference, London, ON, Canada.
12. https://theconversation.com/the-worlds-data-explained-how-much-were-producing-and-where-its-all-stored-159964.
13. www.pge.com/includes/docs/pdfs/myhome/saveenergymoney/homeenergyreport/HER_Beta_SampleReport.pdf.
14. Osama Majeed Butta, Muhammad Zulqarnaina, Tallal Majeed Butt (2021), "Recent advancement in smart grid technology: Future prospects in the electrical power network", Ain Shams Engineering Journal Volume 12, Issue 1, March 2021, pp. 687-695.

3

PVEH based Electrical Energy Creation in Marine Propeller through Computational Hydrodynamic and Vibrational Simulations

Vijayanandh R[1], Naveen Kumar K[1], Raj Kumar G[1], Arul Prakash R[1], Raffik R[2], Kavin R[1] and Rajesh M[1]

[1]*Department of Aeronautical Engineering, Kumaraguru College of Technology, Coimbatore, Tamil Nadu, India*
[2]*Department of Mechatronics Engineering, Kumaraguru College of Technology, Coimbatore, Tamil Nadu, India*

Abstract

A marine propeller is used to provide thrust to ships, boats and unmanned underwater vehicles (UUV). There is always a need for electricity in these vehicles. Thus, to satisfy this, this study presents a method to generate electricity during the mission and arrives at a solution by generating the electric power using a piezoelectric vibrational energy harvester (PVEH). The piezoelectric materials provide the electron movement by the initial impact of the external force, leading to the structural displacements. For the generation of electric power, this work proposes the sticking of a piezoelectric patch over the blades of the propeller. In this regard, a unique propeller is designed for UUV and subsequently modeled in CATIA. The authors have analyzed the marine propellers of different materials but with same dimensions to determine the material which provide the maximum deformation and natural frequencies proportional to the power generation. Also, the power generated by the propeller with different materials is calculated in this work by employing an analytical approach [1]. The computational hydrodynamic and vibrational analyses are major approaches involved in these comparative investigations, where both the computational simulations are analyzed through ANSYS Workbench 17.2. Finally, the suitable lightweight material is proposed to withstand the hydrodynamic environment as well as to extract high electrical energy in marine applications.

3.1 Introduction

Nowadays, there is a growing demand for electric propulsion systems, and our future is seen to be more dependent on electric propulsion. For these systems, the power required by the propulsion system is provided by the batteries in the vehicles, and these batteries play an important role in the overall weight, range and endurance of the vehicles. For many years, to eliminate this problem, researchers have been trying to reduce the weight of the batteries without changing the performance, which in turn can enable the vehicles to carry more payload, thus, enhancing the range and endurance of these vehicles. Another optimum solution is the implementation of an alternate secondary power system that will generate electric power during the mission, as there will always be a need of electricity for this type of propulsion. This secondary power system can power the propulsion system when the primary power system dries out. It can also provide electricity that can charge our mobiles or other important electrical components. It can even provide the voltage required for microprocessors used in communication. The secondary power system will generate electricity from renewable resources. In case of unmanned aerial vehicles (UAV), the use of solar panels for secondary power production is widely considered. In our application, the aim is to generate electric power required for the ships, boats or UUVs as a secondary power system. The piezoelectric materials are suitable for our application and can be stuck over the blades of the marine propeller. The size of the propeller is small when compared to the vehicle it is used for, owing to its working environment. Unlike air, the density of water is high, hence, a small sized propeller is sufficient to provide the required thrust for the vehicle, and a piezoelectric patch over the propeller can provide a significant amount of electric power for the vehicle. In this work, the authors have analyzed the marine propellers of various materials and estimated the amount of power generated by using the piezoelectric material stuck over these.

3.2 Literature Review

Du et al. provided the analytical relationship of piezoelectric patch based electrical energy extraction, verified through the help of MEMS cantilever based electrical test apparatus. Additionally, the authors obtained the optimal design of piezoelectric patches on the cantilever base. The standard beam theory based analytical relationship of the electrical energy was derived with high acceptable outcomes, as the implemented fundamental beam theory was Euler-Bernoulli's principle. From this study, it was observed that the analytical formula of electrical energy primarily relied on design parameters of base set-up, design parameters of piezoelectric patches, forces acting on both base and piezoelectric patch set-ups, natural frequency of base, etc. Thus, in the current study, the investigation of the hydrodynamic forces on the propeller and natural frequency of the marine propeller has been implemented through the computational approaches.

Vijayanandh et al. investigated the hydrodynamic force estimations on marine propeller through the computational and analytical approaches. The hydrodynamic forces such as drag, thrust, lift acting on marine propeller, coefficient of lift and coefficient of drag estimations on marine propeller, torque induced by the marine propeller, mechanical power required to move marine propeller and propulsive efficiency of marine propeller were investigated for hydro-fluid velocities from 5 m/s to 30 m/s. Here, the base engineering approach imposed for the comparative investigations was computational fluid dynamics (CFD) through the computational platform of ANSYS Fluent. The analysis was based on the boundary conditions imposed in hydrodynamic analysis, type of mesh generation and its grid convergence test procedure, type of solver, type of pressure and velocity coupling, analytical formulae involved in the design and performance estimations of marine propeller, etc.

Vijayanandh performed the hydrodynamic force based structural fatigue investigations on the marine propeller through the help of one way coupling based Fluid Structure Interaction approach in ANSYS Workbench. The major lightweight materials used in this investigation were aluminum alloy, stainless steel and unidirectional E-GFRP. The boundary conditions imposed in this work were remote displacement at interior region of the hub, fixed supports at region between outer layer of hub and inner layer of blades, and hydrodynamic load importation on the surface of the marine rotor blades. Here, the first two boundary conditions were analyzed under vibrational analysis. In addition to the materials imposed in this work, two more lightweight composites are used in the current study, which are CFRP and KFRP.

Kumar et al. and Pandian et al. separately modeled and computationally analyzed the unmanned underwater vehicle for various working environments. Both studies dealt with the components and overall weight of the unmanned underwater vehicles, and thereby the final weights of UUVs were estimated, which provided the basic idea to pick the thrust requirement in the current study. The drag generations on entire UUVs were estimated for the velocities ranging from 5 m/s to 15 m/s with the support of the computational hydrodynamic analysis. With the help of these outcomes, the design parameters such as diameter of the propeller, coefficient of lift and coefficient of drag can be estimated easily.

Kulandaiyaappan et al. comprehensively carried out the computational hydrodynamic and computational vibrational analyses over the unmanned aerial vehicle propellers. Through the help of conventional analytical approach focused on the high payload carrying propellers, the design and modeling of ten different aerial propellers were constructed. The propellers were chosen based on high payload capacity, so the design procedures involved in the previous articles were followed in the current study too. Especially, the dielectric constant plays a major role in the analytical estimation of electrical energy through piezoelectric patches, so dielectric constants for lightweight materials for this work are extracted from the previous studies. Also, the analytical procedures followed by Naveen Kumar Kulandaiyaappan et al. support the current investigation.

Vijayanandh Raja et al. comparatively analyzed the hydrodynamic performance factors of a 200 mm diameter based marine propeller. The advanced technique imposed in this investigation was based on a moving reference frame, which provided the platform for the provision of rotational speed to the marine propeller. The transient rotodynamic analyses were carried out from 100 rpm to 10000 rpm on the marine propeller. Consequently, the drag forces, thrust forces, torque induction, requirement of mechanical power and propulsive efficiencies were computed with the help of the computational hydrodynamic analyses in ANSYS Fluent.

The various literature studies provided the ideology for the current study. Especially, the boundary conditions for the computational hydrodynamic and computational vibrational analyses as well as initial conditions such as material properties, turbulence inputs, etc., are predominantly extracted from the literature survey. The major theme of this work is the analytical approach based electrical energy extraction, thus, an analytical relation for this theme has been extracted from validated literature survey [1-3].

3.3 Proposed Methodology

The computational hydrodynamic and computational vibrational analyses are predominantly used in this work to investigate the effect of fluid pressure on hydro-rotor and natural frequency of hydro-rotor. The major focus of the computational hydrodynamic analysis is to estimate the hydrodynamic pressure loads on the picked rotor. The focused outcome of the computational vibrational analysis is to obtain the various natural frequencies of the rotor under the imposition of various lightweight materials and real-time boundary conditions. These two factors represent the major inputs for the analytical approach, as also previously used for the estimation of power extraction through PVEH patch [4-8].

3.3.1 Design of UUV's Propeller

As mentioned earlier, this investigation fully depends upon computational simulation, so the design is the fundamental platform of this work. Therefore, the standard analytical formulae are involved in this investigation.

$$T = 0.5 * \rho * A * [(V_e)^2 - (V_o)^2] \tag{1}$$

The speed of UUV is assumed as 30 knots, which is equal to 15.5 m/s.

$$981 = 0.5 * 1025 * \pi * r^2 * [(15.5)^2 - (2.5)^2]$$

$$\pi * r^2 = 0.0082\ \mathrm{m}^2$$

$$r = 0.051\ \mathrm{m} \approx 50\ \mathrm{mm}$$

Therefore, the diameter of UUV is determined as 100 mm. To estimate the pitch of the UUV's propeller, the historical relationship was executed on pitch and diameter of the marine propellers. The comparative investigations are revealed in Figure 3.1.

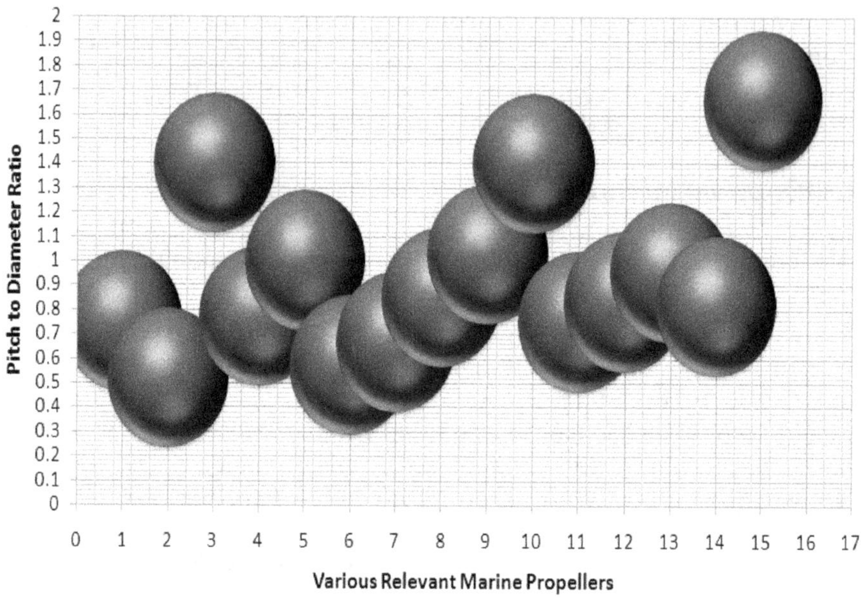

Figure 3.1 The historical relationship between the ratio of pitch and diameter of UUV's propellers.

Through the help of Figure 3.1, the historical relationship is derived, which is given in Equation (2).

$$\frac{\text{Pitch [P]}}{\text{Diameter [D]}} = 0.930807797 \tag{2}$$

$$\text{Pitch [P]} = D * 0.930807797 = 100 * 0.930807797 = 93.1 \text{ mm}$$

Similarly, other two important historical relationships were constructed, which dealt with the pitch and diameter as well as thickness and diameter of various propellers. Figure 3.2 comprehensively reveals the historical relationship between thickness and diameter of UUV's rotor, and consequently a unique relationship formed, which is given in Equation (3).

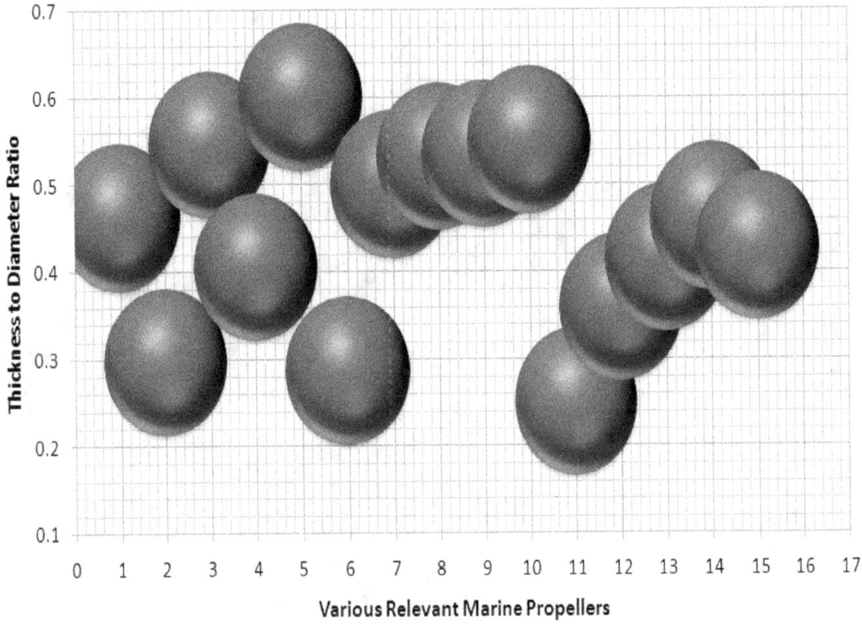

Figure 3.2 The historical relationship between the ratio of thickness and diameter of UUV's propellers.

$$\frac{\text{Thickness [T]}}{\text{Diameter [D]}} = 0.441771695 \quad (3)$$

The third historical relationship is comparatively revealed in Figure 3.3. Here, the predominant factors playing a major role included marine propeller's thickness and marine propeller's pitch [9-13]. The unique connection is formed between thickness and pitch of the rotor, which is exclusively revealed in Equation (4).

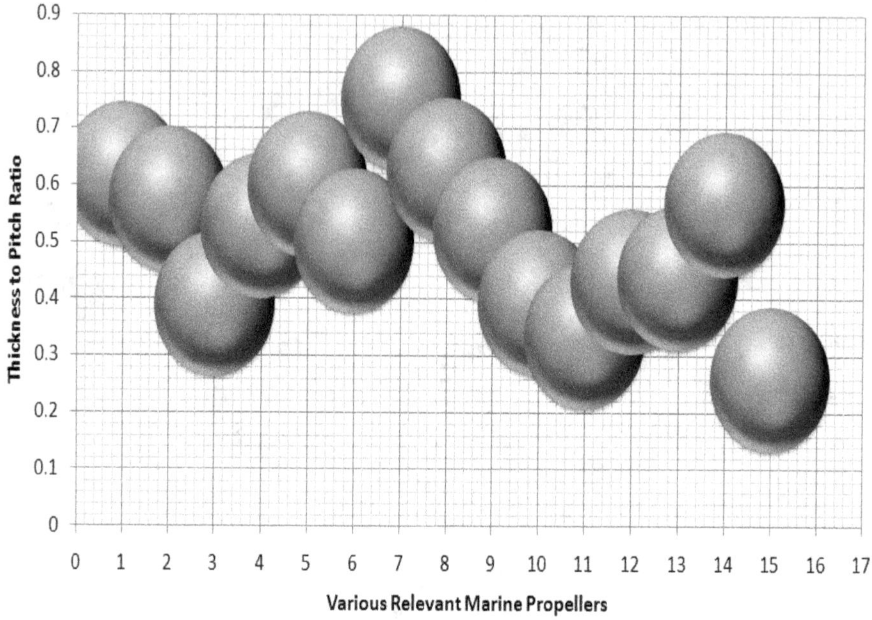

Figure 3.3 The historical relationship between the ratio of thickness and pitch of UUV's propellers.

$$\frac{\text{Thickness [T]}}{\text{Pitch [P]}} = 0.501331315 \tag{4}$$

Through the help of unique Equations (2), (3) and (4), the design of UUV's propeller can be attained for construction. Thus, these historical relationships are a new attempt in the design of the marine propellers. Apart from these Equations (2), (3) and (4), Equations (5) and (6) also played a major role in the design of UUV's propeller. These equations deal with the pitch angle and chord length of the propeller.

$$\theta = \text{arctangent}\left(\frac{P}{2 * \pi * r}\right) \tag{5}$$

$$b = \frac{8 * \pi * \left(\frac{\sin(\theta) * \left(\tan(\theta) - \frac{1}{1.2} * \tan(\theta)\right)}{\left(1 + \frac{1}{1.2} * \tan(\theta)\right)}\right) * r}{n * C_L} \tag{6}$$

The other four Equations (7), (8), (9) and (10) deal with the performance factors of the marine propellers.

$$\text{Advance Coefficient } [J] = \frac{V_i}{N * D} \tag{7}$$

$$\text{Thrust Coefficient } [K_t] = \frac{T}{\rho * N^2 * D^4} \tag{8}$$

$$\text{Torque Coefficient } \left[K_q\right] = \frac{Q}{\rho * N^2 * D^5} \tag{9}$$

$$\text{Propeller Efficiency } [\eta_o] = \frac{T * V_i}{2\pi * N * Q} \tag{10}$$

where, "D" is the overall diameter of the propeller, "N" is the rotational speed of the propeller in rev/sec, "ρ" is the fluid density, "T" is the propeller's thrust, "Q" is the propeller's torque, "Vi" is the inflow velocity, "n" is the number of blades, "C_L" is the coefficient of lift, "θ" is the pitch angle, "P" is the pitch of the propeller, and "b" is the chord length of the propeller [14-20]. The estimated design data of the marine rotor are given in Table 3.1.

Table 3.1 Marine rotor design dimensions

Hub diameter	19 mm
Inner diameter of propeller	20 mm
Outer diameter of propeller	100 mm
Pitch	93.1 mm
Thickness	10 mm
Surface area of each blade	2000 mm^2
Tip to tip length of blade	72 mm

The modeling of the hydro-rotor is completed through CATIA, extracting the design data from Table 3.1. The conceptual design of the fundamental platform is revealed in Figure 3.4. The focused hydro-rotor of this work is to be located in UUV. Based on the UUV's weight and high amount of implementation in real time applications, the number of blades for this work is three. Generally, the UUV lacks endurance, so the alternate electrical energy extraction is very important in hydrodynamic applications.

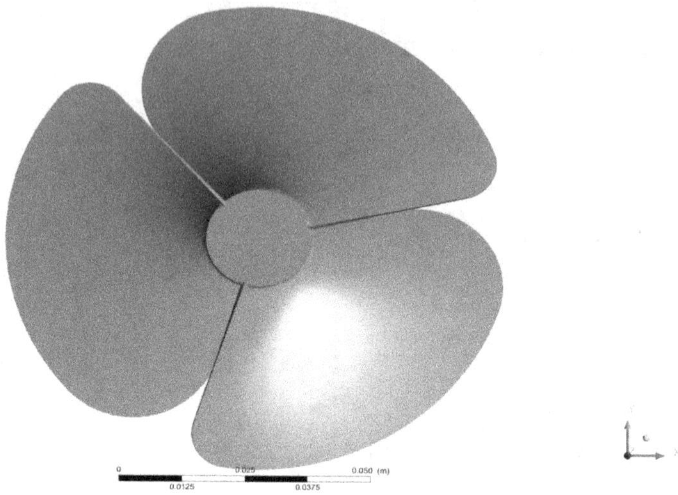

Figure 3.4 The conceptual rotor design.

3.3.2 Mesh

Computationally, mesh is an intermediate process, which converts the normal physical model into a computation model. The computational model has the capability to react with respect to applied loading conditions, whereas the conventional physical model doesn't respond as per the given loads. Thus, the mesh-based conversion is an unavoidable process in the computer-based computations. The fundamental tool for this mesh generation is hydro-rotor, which has a complex curvature design, thus, an unstructured mesh is perfectly suitable methodology to capture each and every curvy path of the rotor. Thus, the unstructured mesh is created for both computational hydrodynamic and vibrational analyses. The discretized structure of computational hydrodynamic computation is revealed in Figure 3.5. Here, the complete naming of the control volume such as MRF (moving reference frame) regions, fixed domain, entry of the flow, exit of the flow and locations of propellers are clearly provided.

3.2.3 Boundary Conditions (BC)

In the computation process, the boundary conditions contribute to the successful execution. This work comprises of two different boundary conditions such as conditions for computational hydrodynamic analysis (CHA) to extract hydrodynamic fluid pressure and conditions for free vibrational analysis to predict mode shape outcomes. In the first case, finite volume approach-based fluid dynamic technique is discussed, wherein a pressure-based solver is used to arrive at the solution, as used for incompressible flows with the working environment also incompressible. The standard k-epsilon based turbulent flow is imposed in the computational fluid dynamic simulation owing to the presence of complex curvature rotor. The fluid is selected to be water with a fluid density of 1025 kg/m^3. The operating hydrodynamic pressure is given as 111380.25 Pa, where the depth of the environment is used as 1 meter, and the gravitational force value is used as 9.81 m/s^2. Under the cell zone condition, the moving reference frame movement is initiated with a given rotational velocity of 100 rpm. The velocity inlet is imposed in this simulation with a velocity of 1 m/s. The pressure outlet is imposed in this computation without any external hydrodynamic pressure. On the propeller, no slip condition is given, and other regions are frictionally expressed with lower behaviors. The second order momentum, second order kinetic energy and second order dissipation rate are imposed with this computation in the SIMPLE based pressure–velocity coupler approach [21-28]. The flow direction and control volumes are expressed in Figure 3.6.

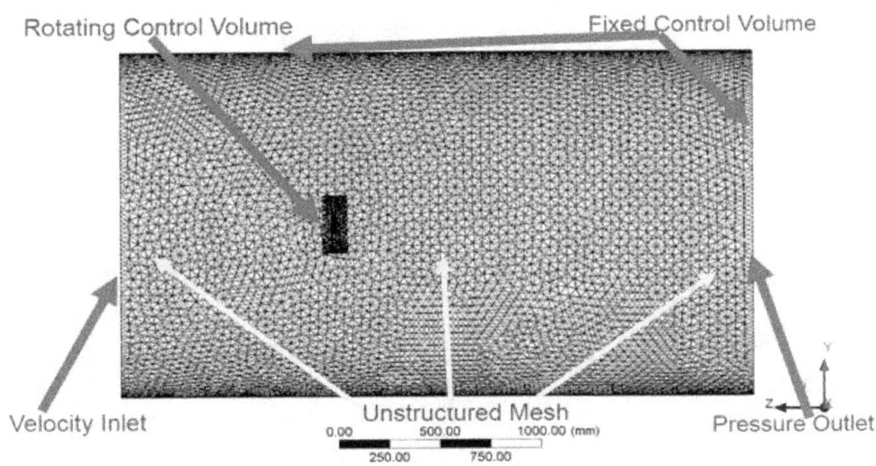

Figure 3.5 The discretized structure of the rotor – advanced fine case.

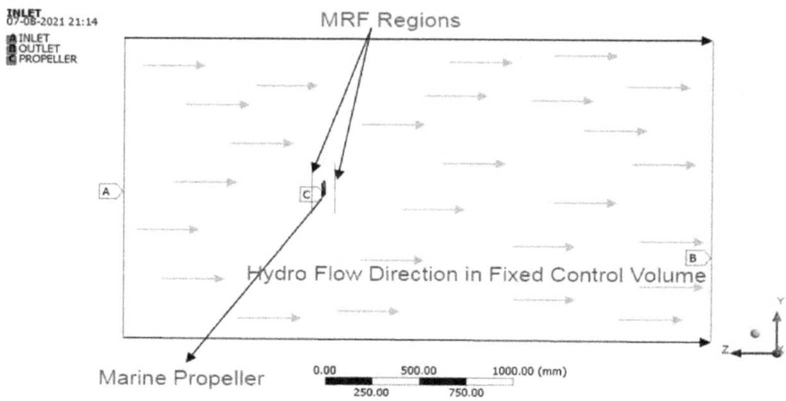

Figure 3.6 The control volume for computational hydrodynamic analysis.

Under the computational vibration analysis (CVA), two major conditions are enforced. These conditions are remote displacement at interior hub and various material properties.

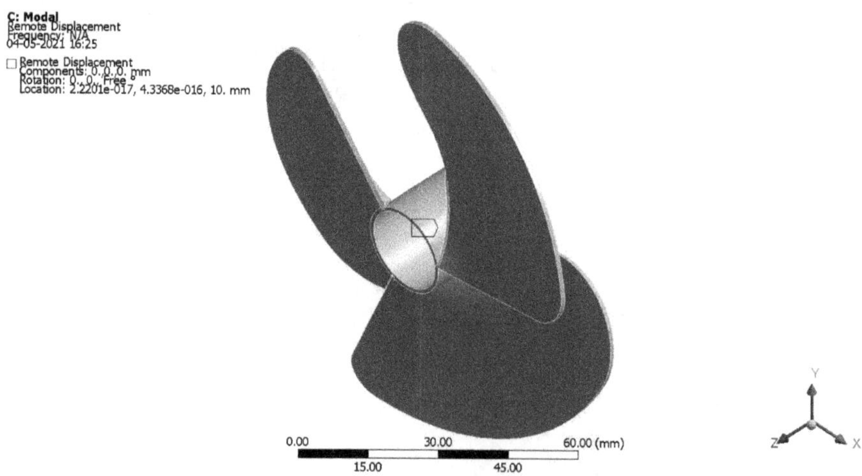

Figure 3.7 The boundary conditions applied on marine propeller under free vibrational case.

As per rotodynamic effect, the marine rotor has the capacity to rotate in axial direction, so the rotation of the marine propeller is expressed perfectly through the remote displacement facility. The lightweight materials imposed in these comparative investigations are CFRP (carbon fiber reinforced polymer)-UD-Prepreg, CFRP-UD-

Wet, CFRP-Woven-Prepreg, CFRP-Woven-Wet, E-GFRP (glass fiber reinforced polymer)-UD, E-GFRP-Wet, GFRP-Woven-FR-4, S-GFRP-UD, KFRP (Kevlar fiber reinforced polymer)-49-UD, aluminum alloy, polyethylene, magnesium alloy, stainless steel, gray cast iron and titanium alloy. The material properties are collected from previous published articles [1–46] and imposed in these comparative investigations.

3.2.4 Grid Independence Studies

Normally, the grid convergence tests are carried out in order to choose the suitable mesh elements, which are able to provide acceptable cum reliable outcomes. Thus, this investigation implemented two different grid convergences, which comprised of grid convergence test on hydrodynamic pressure and grid convergence test on natural frequency. Figures 3.8 and 3.9 correspond to the first grid convergences test. From Figure 3.8, it is observed that mesh case – 3 is the best to provide the aforesaid computational outcomes. Therefore, the same mesh case – 3 is followed in the main computational hydrodynamic simulation. Figure 3.9 reveals the mesh case – 3, which is unstructured with fine curvature and area proximity-based discretization.

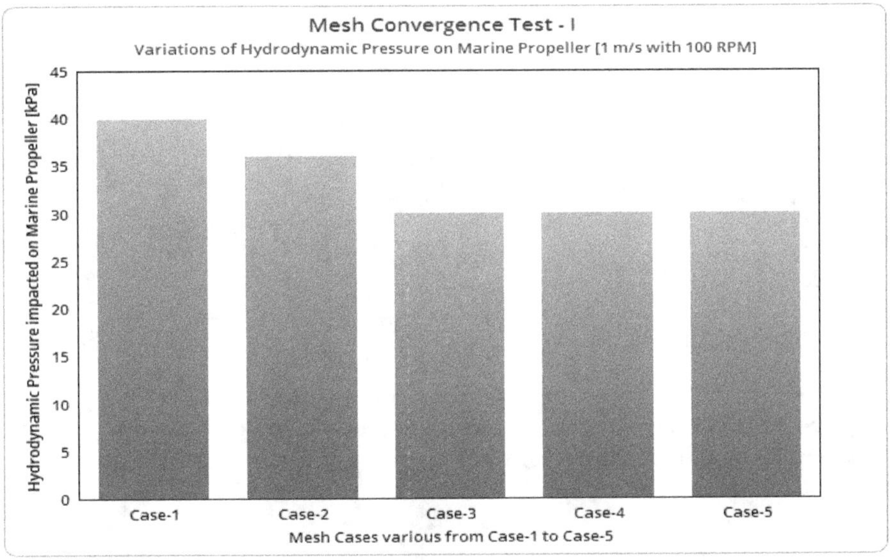

Figure 3.8 The first grid independent study on hydrodynamic pressure.

Figure 3.9 The mesh case – 3 (sectional views of all the control volumes).

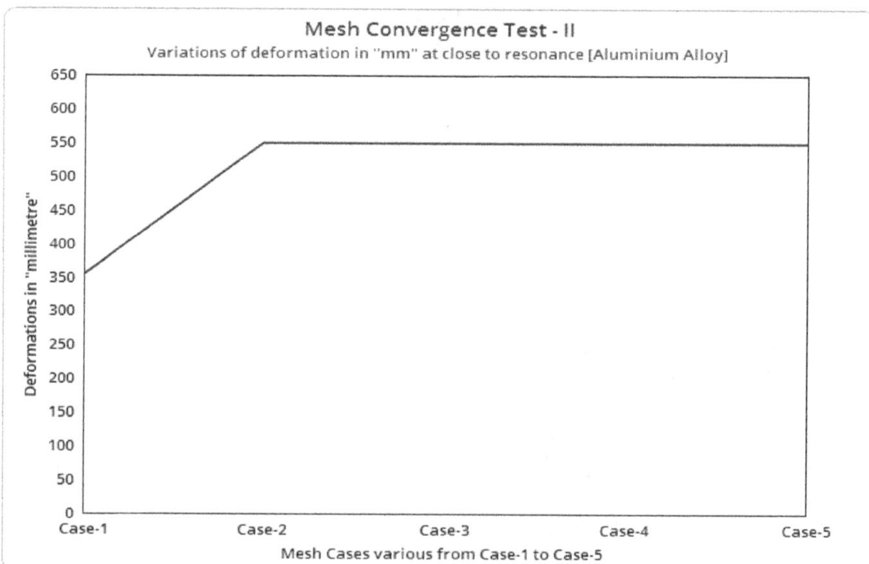

Figure 3.10 The second grid independent study on deformation of aluminum alloy under free vibration.

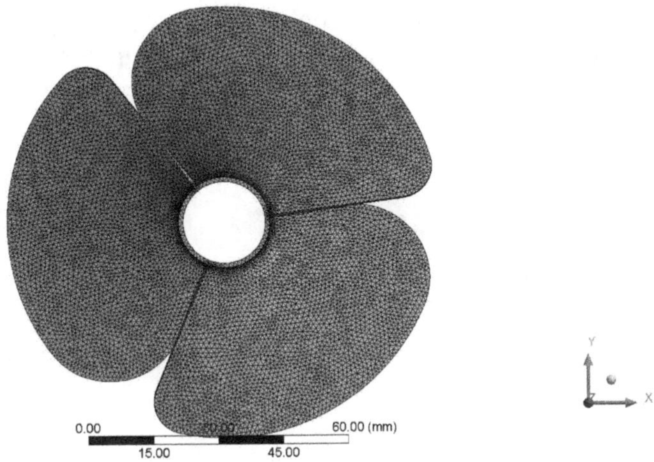

Figure 3.11 The discretized structure of the rotor – medium fine case (case – 2).

The second convergence test-based outcomes are executed, and the outcomes are shown in Figures 3.10 and 3.11. The comprehensive outcome of the second grid convergence test is projected in Figure 3.10, and the shortlisted mesh case – 2 is revealed in Figure 3.11. For this test, the aluminum alloy and its natural frequencies are picked as tester, where the fifth mode is chosen as the suitable displacement provider [29-46].

3.4 Results and Discussion

The CHA and CVA computations are executed as per the aforesaid boundary conditions, and the results are monitored. Both CHA and CVA computations have been worked under the principles of various computational procedures such as finite volume approach and finite element approach. Specifically, CHA is computed based on the finite volume technique, and CVA is computed based on the 3-D finite element method. The major outcomes of CHA and CVA are the predominant inputs to the analytical estimations of electricity generation through the PVEH patches.

3.4.1 Hydrodynamic Results

Figures 3.12 and 3.13 provide the hydrodynamic results of the UUV propeller under the normal working conditions. The hydrodynamic pressure load on the propeller is the major input to the PVEH electricity generation, where the relationship directly relates with the voltage generation. Thus, CHA is unavoidable cum mandatory one.

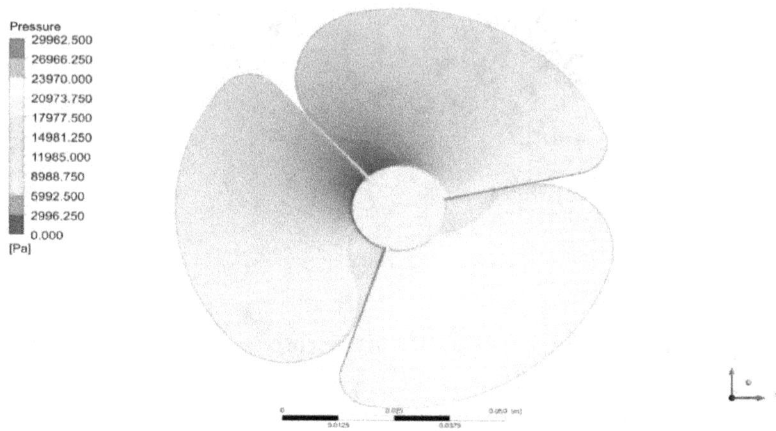

Figure 3.12 The hydrodynamic pressure variations on hydro-rotor.

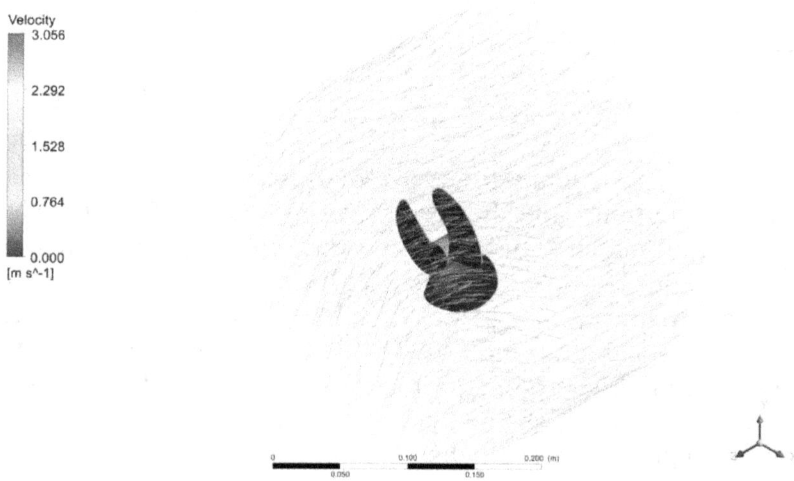

Figure 3.13 The hydro-fluid distribution over the hydro-rotor.

3.4.2 Free Vibrational Results

The natural frequency of the structure is found out and is subsequently used. When the propeller is operated at natural frequency, it will break but when operated at a frequency less than it, the displacement will be maximum which will yield us the maximum output power. Also, the propeller should be maintained within the elastic limit as there will be permanent deformation in case it crosses the limit, which is undesirable. These conditions will enable the UUV propeller to generate energy without breaking. Thus, free vibrational analyses are carried out on the rotor with aforesaid boundary conditions for various lightweight materials. The natural frequency based computational outcomes are revealed in Figures 3.14 to 3.18. In the free vibrational analysis, mode shape and its outcomes such as deformations and natural frequencies are involved in the selection. Thereafter, a suitable mode shape is picked, which is fifth mode shape based on high deformation and natural frequency. Figure 3.14 reveals the outcome of aluminum alloy, whereas the behavior of the CFRP UD Prepreg based composite is revealed in Figure 3.15. Further, E-Glass Wet based composite is revealed in Figure 3.16, KFRP-UD-49 based composite is revealed in Figure 3.17, and finally polyethylene relevant outcome is shown in Figure 3.18.

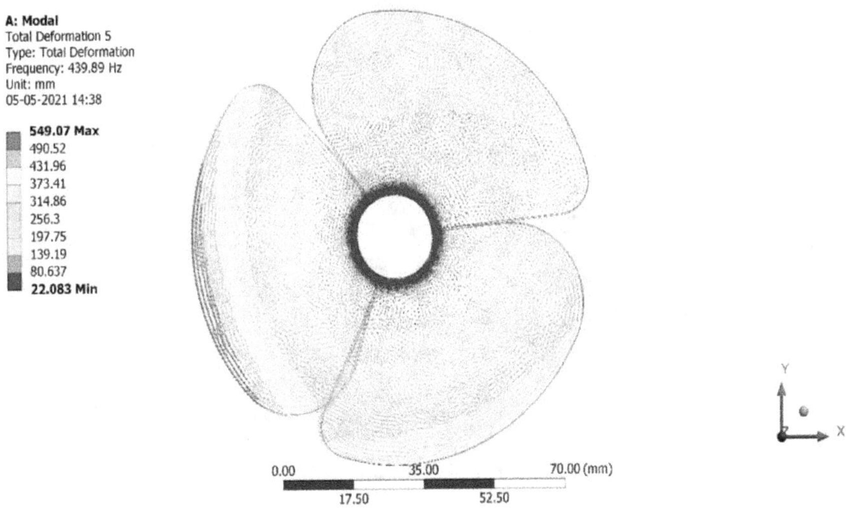

Figure 3.14 The fifth mode shape outcome of aluminum alloy.

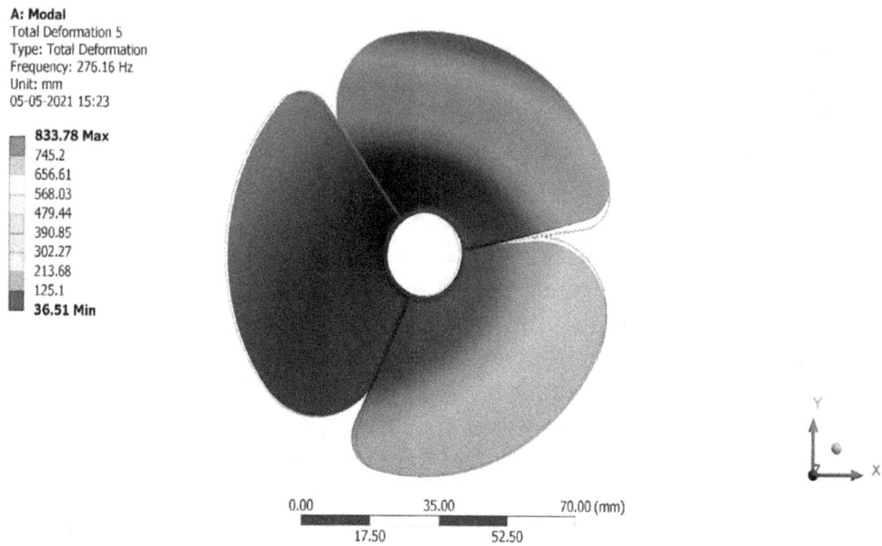

Figure 3.15 The fifth mode shape outcome of CFRP UD prepreg-based composite.

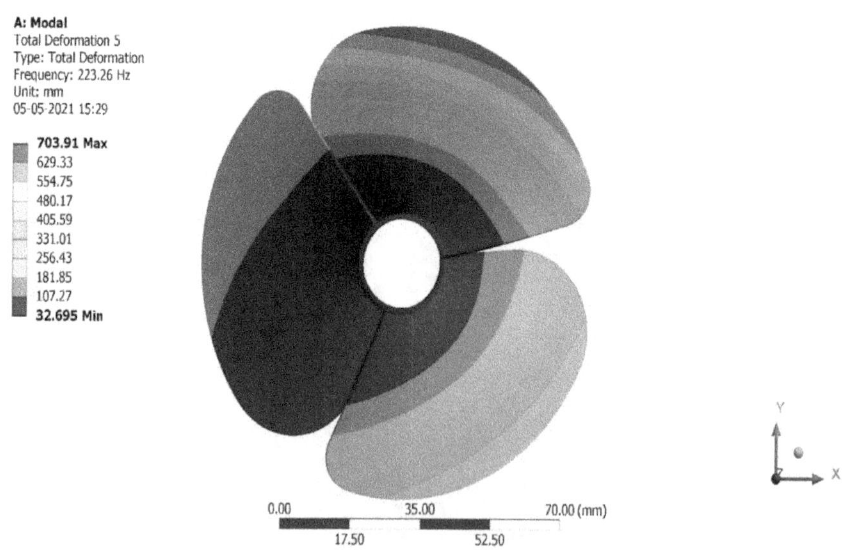

Figure 3.16 The fifth mode shape outcome of E-glass wet based composite.

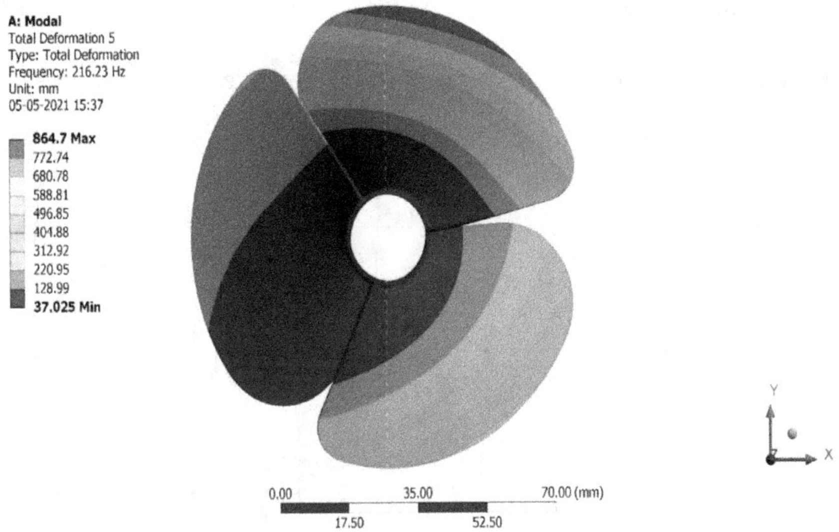

Figure 3.17 The fifth mode shape outcome of KFRP-UD-49 based composite.

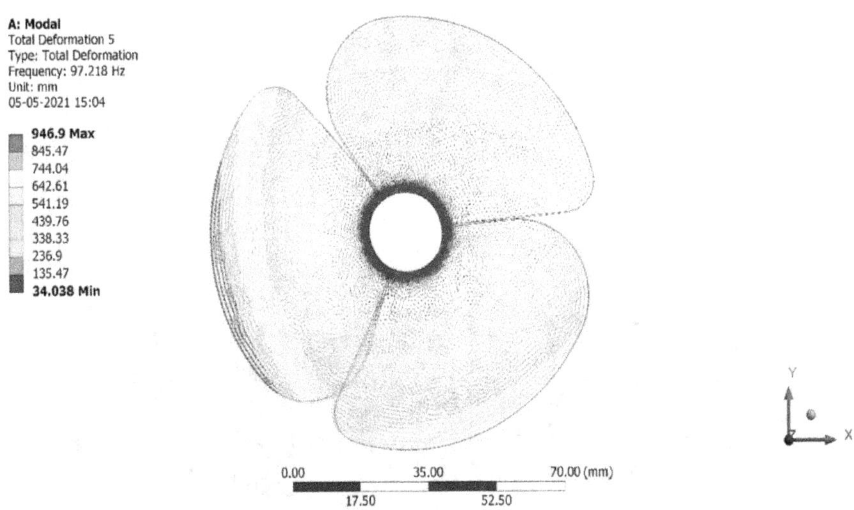

Figure 3.18 The fifth mode shape outcome of polyethylene.

3.4.3 Comparative Analysis

Figures 3.19 and 3.20 contain the comprehensive natural frequency outcomes of the various lightweight materials.

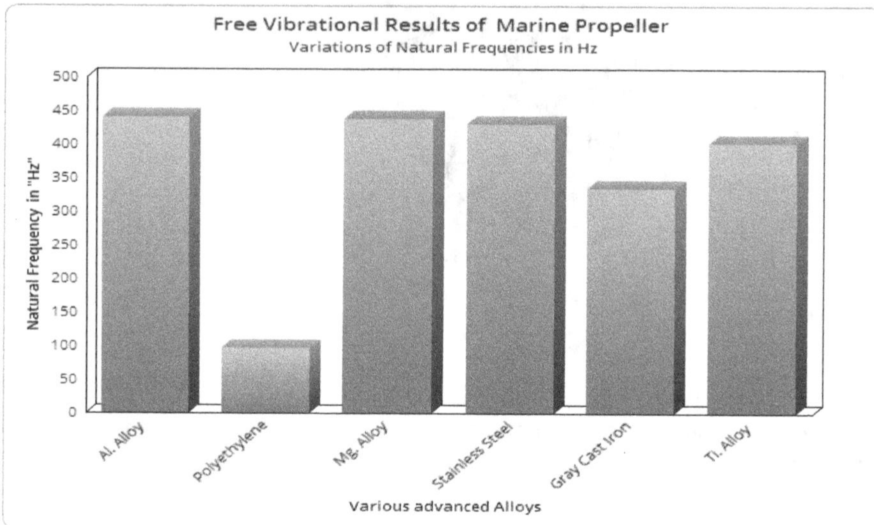

Figure 3.19 The free vibrational results of alloy-based marine propeller.

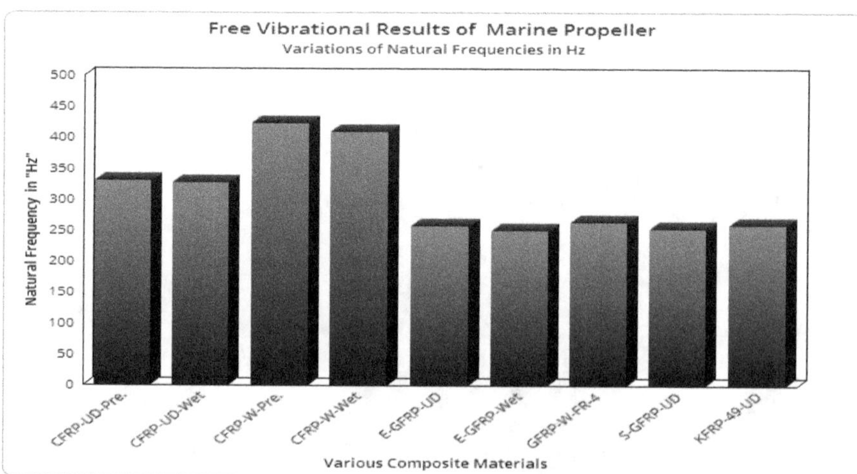

Figure 3.20 The free vibrational results of the various composites based marine propellers.

Here, Figure 3.19 reveals the natural frequency outcomes of the lightweight alloys, and Figure 3.20 presents the natural frequency outcomes of the lightweight composites.

3.5 PVEH Based Electricity Estimation

The proposed alternate energy extraction in this investigation is pictorially represented and revealed in Figures 3.21 and 3.22.

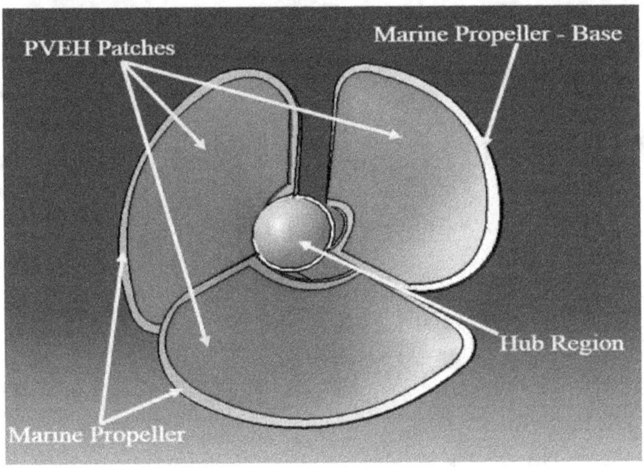

Figure 3.21 The PVEH patches-based hydro-rotor – front side view.

Equations (11) and (12) contain the compositional analytical elements of electricity generation through PVEH patches, as derived and discussed in the literature survey [1]. In these analytical formulae, three important inputs are obtained from the computational simulations and/or literature survey. These inputs include: hydrodynamic pressure based uniformly distributed loads, natural frequencies and dielectric constants. Here, the hydrodynamic pressure and natural frequencies are computed through CHA and CVA respectively, and the dielectric constants are obtained from the literature [1-41].

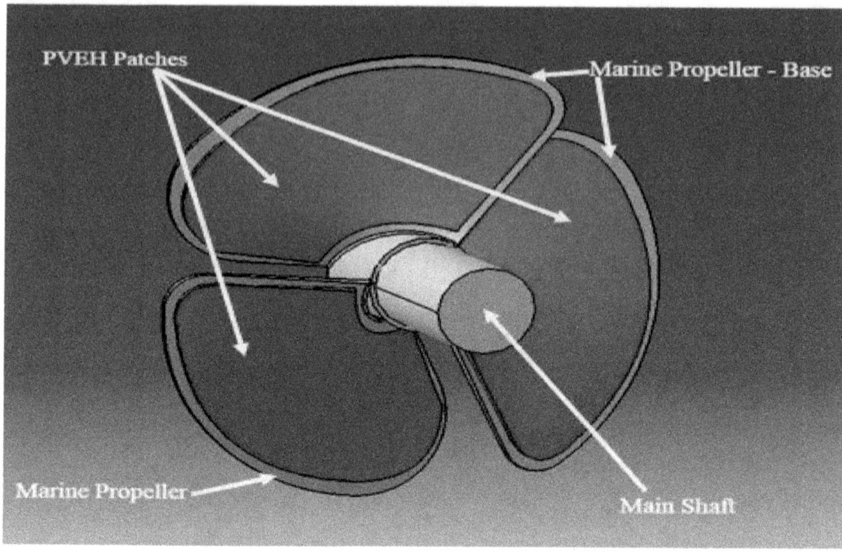

Figure 3.22 The PVEH patches-based hydro-rotor – bottom view.

$$P_{Intermediate} = d_{pmc}^2 w^2 f^2 * \frac{18 * T_{PL}}{[(W_P) * (L_{PL})^2 * ([t_P] + [T_{PL}])^4]} * \frac{\rho_{lmw}}{[1 + f * \varepsilon * (\rho_{lmw})]} \tag{11}$$

$$P_{Final} = (P_{Intermediate}) * \left(\frac{\ell^5}{36} - \frac{L\ell^4}{6} + \frac{5L^2\ell^3}{12} - \frac{L^3\ell^2}{2} + \frac{L^4\ell}{4}\right) \tag{12}$$

where, "d_{pmc}" is the piezoelectric material constant (0.12), "w" is the aerodynamic load (N), "f" is the natural frequency (Hz), "W_P" is the width of the UUV's propeller (m), "L_{PL}" is the length of the piezoelectric layer (m), "t_P" is the thickness of the wing (m), "T_{PL}" is the thickness of the piezoelectric layer (m), "ρ_{lmw}" is the density of the material, and "ε" is the permittivity of the materials. For this work, the sample calculation is taken and calculated as CFRP-Woven-Wet,

$$P_{\text{Intermediate}} = d_{\text{pmc}}^2 w^2 f^2 * \frac{18 * T_{PL}}{[(W_P) * (L_{PL})^2 * ([t_P] + [T_{PL}])^4]} * \frac{\rho_{lmw}}{[1 + f * \varepsilon * (\rho_{lmw})]}$$

The input values are: piezoelectric material constant (d_{pmc}) = 0.12; area of the UAV's rotor (single side of single leaf] = 0.0024806 m^2; aerodynamic pressure on UUV's propeller = 29962.5 N/m^2; force (w) (single side of single leaf) = 29962.5 * 0.0024806 = 74.3249775 N; natural frequency (f) = 409.52 Hz; width of the UUV propeller (W_P) = 40.708 mm = 0.040708 m; length of the piezoelectric layer (L_{PL}) = 72.097 mm = 0.072097 m; thickness of the UUV propeller (t_P) = 0.99997 mm = 0.00099997 m; thickness of the piezoelectric layer (T_{PL}) = 0.99997 mm = 0.00099997 m; density of the material (ρ_{lmw}) = 1451kg/m^3; and $\varepsilon_r \varepsilon_0 = \varepsilon =$ permittivity of the material = dielectric constant = 5;

$$P_{\text{Intermediate}} = 1435.459533 \text{ Watts m}^5$$

Thereby,

$$P_{\text{Final}} = (P_{\text{Intermediate}}) * \left(\frac{\ell^5}{36} - \frac{L\ell^4}{6} + \frac{5L^2\ell^3}{12} - \frac{L^3\ell^2}{2} + \frac{L^4\ell}{4}\right)$$

$$\text{Electrode length } (\ell) = 0.44 * 0.072097 = 0.03172268 \text{ m}$$

$$P_{\text{Final}} = 119.972 \ \mu\text{W}$$

Similarly, this calculation is extended for all other fourteen lightweight materials, and the electricity outcomes are revealed in Figures 3.23 to 3.26.

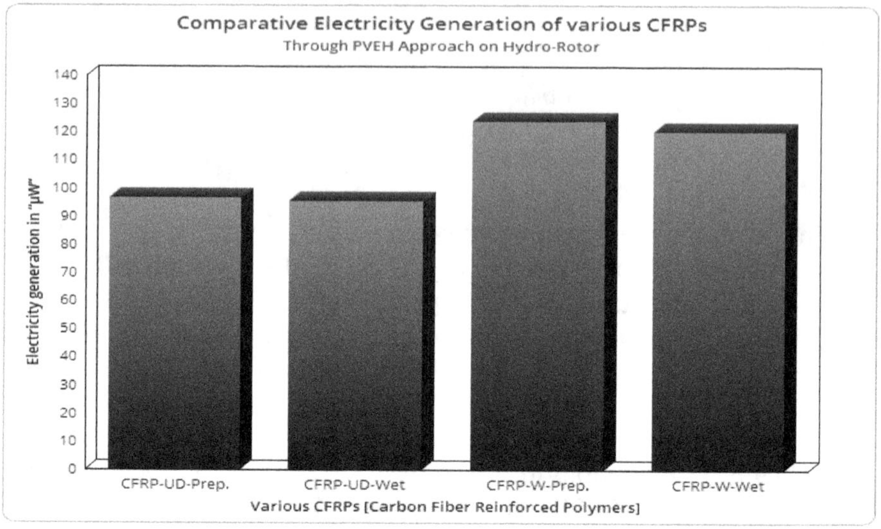

Figure 3.23 The comprehensive electricity generation level of various CFRPs.

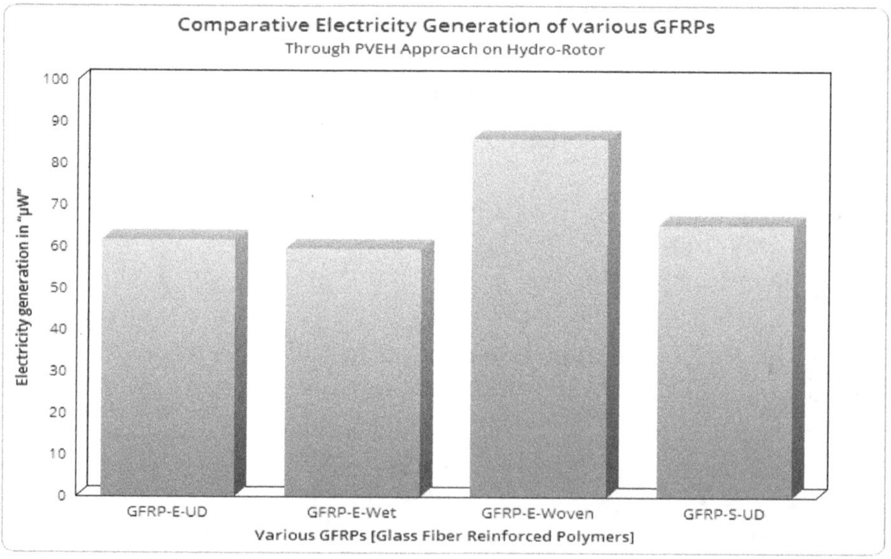

Figure 3.24 The comprehensive electricity generation level of various GFRPs.

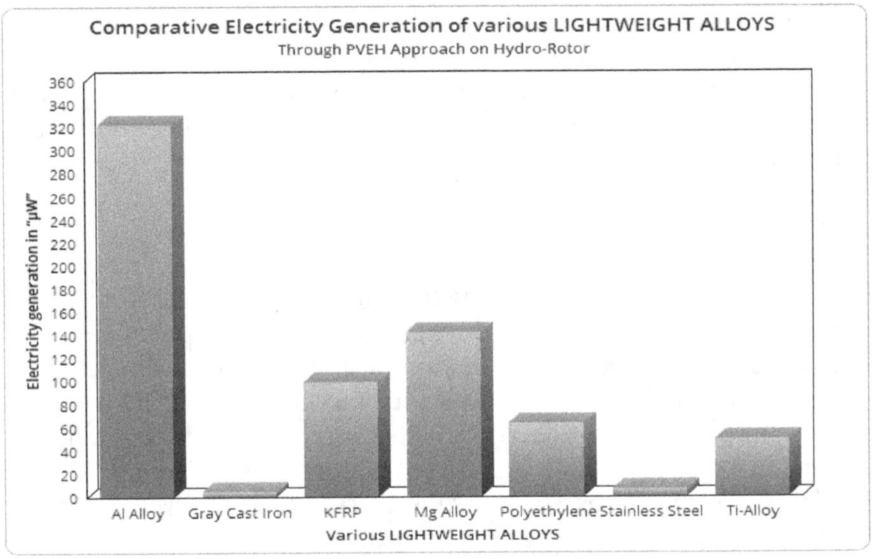

Figure 3.25 The comprehensive electricity generation level of various lightweight materials.

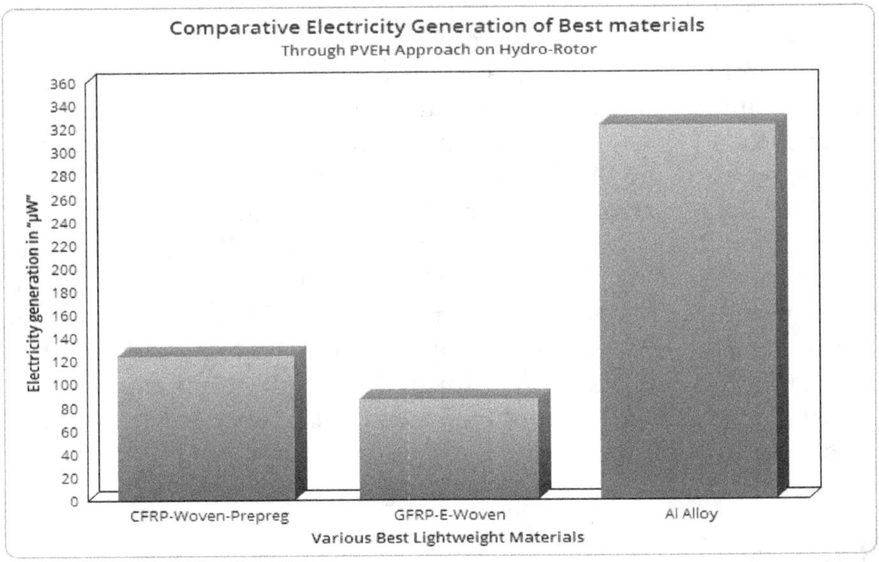

Figure 3.26 The comprehensive electricity generation level of various lightweight materials.

From Figures 3.23 to 3.26, it is strongly observed that the aluminum alloy performed better than other lightweight materials owing to its high vibration. Due to this, the aluminum alloy generated higher electricity, thus, this material is suggested to be implemented in the real time applications.

3.5 Conclusions

The historical relationships between the thickness and pitch, pitch and diameter, thickness and diameter of UUV's propeller are collected, and the unique relationships are subsequently developed. Based on the formed relationships and standard analytical formulae, the design parameters of the propeller are estimated. Through the help of the modeling tool, i.e., CATIA, the conceptual model of the UUV's propeller is constructed. The unstructured discretization on the UUV's propeller and two grid convergence tests are conducted to confirm the suitable mesh cases. The optimized mesh case supported the execution of computations such as CHA and CVA in ANSYS Workbench 17.2. Through CHA, the hydrodynamic pressures and hydrodynamic loads are estimated on UUV's propeller, which is predominant input for electricity generation. Next, the various lightweight materials-based propellers are imposed in CVA computations, and the corresponding displacements and natural frequencies are tabulated for further execution. With the help of the analytical approach [1], the electrical energy generation in propellers are calculated through PVEH patches. From the comparison Figures 3.23 to 3.26, it is strongly observed that aluminum alloy performed better than other lightweight materials owing to its high vibration. Due to this, the alloy generated higher electricity, thus, this material is suggested for the real time applications. Additionally, based on the integrative effects, the authors propose that the woven fiber based CFRP composites also fit for marine applications as these can provide a reasonable amount of electrical energy extraction and high structural lifetime.

References

1. Sijun Du, Yu Jia and Ashwin Seshia, (2015) "Maximizing Output Power in a Cantilevered Piezoelectric Vibration Energy Harvester by Electrode Design", Journal of Physics: Conference Series, 660 012114, pp. 1-5, DOI:10.1088/1742-6596/660/1/012114.

2. R Vijayanandh, K Venkatesan, R Raj Kumar, M Senthil Kumar, G Raj Kumar and P. Jagadeeshwaran, (2020), "Theoretical and Numerical Analyses on Propulsive Efficiency of Unmanned Aquatic Vehicle's Propeller", IOP Journal of Physics: Conference Series, 1504 012004, pp. 1-10, https://doi.org/10.1088/1742-6596/1504/1/012004.
3. Vijayanandh R, Venkatesan K, Senthil Kumar M, Raj Kumar G, P. Jagadeeshwaran, Raj Kumar R, (2020) "Comparative fatigue life estimations of Marine Propeller by using FSI", IOP - Journal of Physics: Conference Series, 1473, 012018, pp. 1-8.
4. Naveen Kumar Kulandaiyaappan et al., (2021) 'Optimization of High Payload Unmanned Aerial Vehicle's Propellers based on Energy Formation by using Computational Vibrational Analyses', AIAA Propulsion and Energy 2021 Forum, VIRTUAL EVENT, https://doi.org/10.2514/6.2021-3729.
5. Vijayanandh Raja et al., (2021) 'Comparative Estimations of Hydrodynamic Analysis on Unmanned Aquatic Vehicle's Propeller by using an advanced [CFD with MRF] Approach', AIAA Propulsion and Energy 2021 Forum, VIRTUAL EVENT, https://doi.org/10.2514/6.2021-3732.
6. V Praveen Kumar, S Kishor Kumar, K R Sankaresh Pandian, E Ashraf, K Thanga Tamil Selvan, Vijayanandh R, (2019) "Conceptual Design And Hydrodynamic Research On Unmanned Aquatic Vehicle", International Journal of Innovative Technology and Exploring Engineering, Vol. 8, Issue 11S, pp 121-127, DOI: 10.35940/ijitee.K1027.09811S19.
7. Vijayanandh R, Senthil Kumar M, Rahul S, Thamizhanbu E, (2020) "Conceptual Design and Comparative CFD Analyses on Unmanned Amphibious Vehicle for Crack Detection", Lecture Notes in Civil Engineering, 14, pp. 133-150, https://doi.org/10.1007/978-3-030-37393-1_14.
8. Nneka Osuchukwu, Leonid Shpanin, (2017) "Feasibility Study of the Quadcopter Propeller Vibrations for the Energy Product, International Journal of Mechanical and Mechtronics Engineering, Vol:11, No:2, pp. 1-7, DOI:10.5281/zenodo.1340064.
9. Steven R. Anton, Daniel J. Inman, (2011), Performance modeling of unmanned aerial vehicles with on-board energy harvesting, Proceedings Volume 7977, Active and Passive Smart Structures and Integrated Systems 2011; 79771H, https://doi.org/10.1117/12.880473.
10. Steven R. Anton, Alper Erturk, Daniel J. Inman, (2012) "Investigation of a Multifunctional energy harvesting and energy-storage wing spar for unmanned aerial vehicles", Journal of Aircraft, Vol. 49, No. 1, pp. 292-301, DOI: 10.2514/1.C031542.

11. Saber Mohammadi, Kaveh Cheraghi and Akram Khodayari, (2019) Piezoelectric vibration energy harvesting using strain energy method, Engineering Research Express, Volume 1, Number 1, https://orcid.org/0000-0002-7950-3027.
12. Robert Kress and E. L. Lorenz, Marine Propeller Selection, SAE Transactions, Vol. 79, Section 1: Works 700010–700182 (1970), pp. 334-345 (12 pages).
13. Sushree S Sahoo, Vijay K Singh and Subrata K Panda, (2015) "Experimental and simulation study of flexural behaviour of woven Glass/Epoxy laminated composite plate", IOP Conf. Series: Materials Science and Engineering, 75 012017, pp. 1 - 10, DOI:10.1088/1757-899X/75/1/012017.
14. Saber Mohammadi, Kaveh Cheraghi and Akram Khodayari, (2019) "Piezoelectric vibration energy harvesting using strain energy method", Eng. Res. Express, 1, 015033, pp. 1-16, DOI:10.1088/2631-8695/ab3f0c.
15. M. Senthil Kumar, Vijayanandh R, (2018) "Vibrational fatigue analysis of NACA 63215 small horizontal axis wind turbine blade", Materials Today Proceedings, Volume 5, Issue 2, Part 2, Pages 6665–6674, https://doi.org/10.1016/j.matpr.2017.11.323.
16. Hassan Elahi, Marco Eugeni and Paolo Gaudenzi, (2018) 'A Review on Mechanisms for Piezoelectric-Based Energy Harvesters, Energies', 11(7),1850, pp. 1-35, DOI:10.3390/en11071850.
17. Carlos De Marqui, Wander G. R. Vieira and Daniel J. Inman, (2011) "Modeling and Analysis of Piezoelectric Energy Harvesting from Aeroelastic Vibrations Using the Doublet-Lattice Method", Journal of Vibration and Acoustics, Vol. 133 / 011003, pp. 1-9, DOI: 10.1115/1.4002785.
18. Steven Dobbs, Zhen Yn, Kevin R. Anderson, Jonathan A. Franco, Alexander E. Deravanersian, Albert Lin, Andrew Ahn, (2018) "Design of an In-flight Power Generation and Storage System for UAVs", IEEE Conference on Technologies for Sustainability, Conference Series 978-1-5386-7791-9/18, pp. 1-8, DOI:10.1109/Sus Tech.2018.8671363.
19. P. Jagadeeshwaran, V. Natarajan, Vijayanandh R, Senthil Kumar M, Raj Kumar G, (2020) "Numerical Estimation of Ultimate Specification of Advanced Multi-Rotor Unmanned Aerial Vehicle", International Journal of Scientific & Technology Research, Vol. 9, Issue 01, pp. 3681-3687.
20. Raj Kumar G, Vijayanandh R, Senthil Kumar M, Naveen Kumar K, Ahilla Bharathy L, (2017) "Conceptual design and structural analysis of integrated composite Micro Aerial Vehicle", Journal of Advanced Research in Dynamical and Control Systems, Vol. 9. Sp– 14, pp 857-881.

21. Vijayanandh R., Darshan Kumar J., Senthil Kumar M., Ahilla Bharathy L., Raj Kumar G. (2019) "Design and Fabrication of Solar Powered Unmanned Aerial Vehicle for Border Surveillance", Springer Series in Geomechanics and Geoengineering, 7, pp. 61-71, https://doi.org/10.1007/978-3-319-77276-9_7.
22. Vijayanandh. R, Ramesh. M, Senthil Kumar. S, Raj Kumar. G, Senthil Kumar. M & Naveen Kumar. R, 2018 The Conceptual Design of the Tilt-Copter Based on the Speech Control (A Theoretical Approach), International Journal of Mechanical and Production Engineering Research and Development, Vol. 8, Spl. 7, 423-439.
23. Vijayanandh R, Ramesh M, Raj Kumar G, Thianesh U K, Venkatesan K, Senthil Kumar M, (2019) 'Research of Noise in the Unmanned Aerial Vehicle's Propeller using CFD', International Journal of Engineering and Advanced Technology, ISSN: 2249 – 8958, Vol. 8, Issue 6S, pp 145 – 150, DOI:10.35940/ijeat. F1031.0886S19.
24. Senthil Kumar M, Vijayanandh R & Gopi B, (2018), "Numerical Investigation on Vibration Reduction in Helicopter Main Rotor Using Air Blown Blades", International Journal of Mechanical and Production Engineering Research and Development, Vol. 8, Special Issue 7, pp. 152-164.
25. Balaji S, Prabhagaran P, Vijayanandh R, Senthil Kumar M, Raj Kumar R, (2020) "Comparative Computational Analysis on High Stable Hexacopter for Long Range Applications", Lecture Notes in Civil Engineering, 31, pp. 369-391, https://doi.org/10.1007/978-3-030-37393-1_31.
26. Vijayanandh R, Kiran P, Indira Prasanth S, Raj Kumar G, Balaji S, (2020) "Conceptual Design and Optimization of Flexible Landing Gear for Tilt-Hexacopter Using CFD", Lecture Notes in Civil Engineering, 15, pp. 151-174, https://doi.org/10.1007/978-3-030-37393-1_15.
27. R. Vijayanandh, M. Ramesh, K. Venkatesan, M. Senthil Kumar, and R. Rajkumar, (2021) "Comparative Acoustic Analysis of Modified Unmanned Aerial Vehicle's Propeller", Lecture Notes in Mechanical Engineering, 45, pp. 557-571, https://doi.org/10.1007/978-981-15-5996-9_45.
28. Balaji Sonaimuthu, Prabhagaran Panchalingam, Vijayanandh Raja, (2019) "Comparative Analysis of Propulsive System in Multi-Rotor Unmanned Aerial Vehicle", Proceedings of the ASME, Gas Turbine India Conference - GTINDIA2019, V002T08A004, 2019, pp. 1-8, https://doi.org/10.1115/GTINDIA2019-2429.
29. Vijayanandh R, Senthil Kumar M, Vasantharaj C, Raj Kumar G, Soundarya S, (2017) "Numerical Study on Structural Health Monitoring for Unmanned Aerial Vehicle", Journal of Advanced Research in Dynamical and Control Systems, Vol. 9. Sp– 6, pp 1937 – 1958.

30. Ramesh Murugesan, Vijayanandh Raja, (2017) "Acoustic Investigation on Unmanned Aerial Vehicle's Rotor Using CFD-MRF Approach", Proceedings of the ASME 2019, Gas Turbine India Conference - GTINDIA2019, V002T08A005, pp. 1-7, https://doi.org/10.1115/GTINDIA2019-2430.
31. Vijayanandh R, Naveen Kumar K, Senthil Kumar M, Raj Kumar G, Naveen Kumar R, Ahilla Bharathy L, (2018) "Material Optimization of High-Speed Micro Aerial Vehicle using FSI Simulation", Procedia Computer Science, ISSN 1877-0509, Vol. 133, pp 2-9, https://doi.org/10.1016/j.procs.2018.07.002.
32. Vijayanandh R, M. Senthil Kumar, K. Naveenkumar, G. Raj Kumar, R. Naveen Kumar, (2017) "Design Optimization of Advanced Multirotor Unmanned Aircraft System Using FSI", Lecture Notes in Mechanical Engineering, 28, pp. 299-310, DOI 10.1007/978-981-13-2718-6.
33. K. Naveen Kumar, R. Vijayanandh, G. Raj Kumar, B. Sanjeev, Hariharan Balachander and S. Guru Prasad, (2018) "Comparative Approaches for Fatigue Life Estimation of Aluminium Alloy for Aerospace Applications", Int. J. Vehicle Structures & Systems, 10, 4, pp. 282-286, DOI: 10.4273/ijvss.10.4.11.
34. S. Bhagavathiyappan, M. Balamurugan, M. Rajamanickam, R. Vijayanandh, G. Raj Kumar, and M. Senthil Kumar, (2020) "Comparative computational impact analysis of multi-layer composite materials", AIP Conference Proceedings, 2270, pp. 040007-1 to 040007-5, https://doi.org/10.1063/5.0019380.
35. Raj Kumar. G, Senthil Kumar. M, Vijayanandh. R, K. Raja Sekar, Mohamed Bak. K &Varun. S, (2019) 'The Mechanical Characterization of Carbon Fiber Reinforced Epoxy with Carbon Nanotubes', International Journal of Mechanical and Production Engineering Research and Development, ISSN(E): 2249-8001, Vol. 9, Special Issue 1, 243-255.
36. Vijayanandh R, Venkatesan K, Ramesh M, Raj Kumar G, Senthil Kumar M, (2019) 'Optimization of Orientation of Carbon Fiber Reinforced Polymer Based on Structural Analysis', International Journal of Scientific & Technology Research, ISSN 2277-8616, Volume 8 - Issue 11, 3020 – 3029.
37. Naveen Kumar K, Vijayanandh R, Bruce Ralphin Rose J, Swathi V, Narmatha R, Venkatesan. K, (2019) "Research on Structural behavior of Composite Materials on different Cantilever Structures using FSI", International Journal of Engineering and Advanced Technology, Vol. 8, Issue 6S3, pp. 1075-1086, DOI: 10.35940/ijeat.F1178.0986S319.
38. S. Indira Prasanth, K. Kesavan, P. Kiran, M. Sivaguru, R. Sudharsan, and R. Vijayanandh, (2020) "Advanced structural analysis on E-glass fiber reinforced with polymer for enhancing the mechanical

properties by optimizing the orientation of fiber", AIP Conference Proceedings, 2270, pp. 040006-1 to 040006-5, https://doi.org/10.1063/5.0019378.

39. Naveen Kumar K, Vijayanandh R, Bruce Ralphin Rose J, Swathi V, Narmatha R, Venkatesan. K, (2019) Research on Structural behavior of Composite Materials on different Cantilever Structures using FSI, International Journal of Engineering and Advanced Technology, 8, 6S3, pp: 1075 - 1086, DOI: 10.35940/ijeat.F1178.0986S319.
40. Rajagurunathan. M, Raj Kumar. G, Vijayanandh. R, Vishnu. V, Rakesh Kumar. C & Mohamed Bak. K, (2018) The Design Optimization of the Circular Piezoelectric Bimorph Actuators Using FEA, International Journal of Mechanical and Production Engineering Research and Development, 8, 7, 410-422.
41. R. Sankaresh Pandian, R. Vijayanandh, S. Kishor Kumar, V. Praveen Kumar, M. Ramesh, M. Senthil Kumar, and G. Raj Kumar, (2021) "Comparative Hydrodynamic Investigations on Unmanned Aquatic Vehicle for Ocean Applications", Recent Trends in Mechanical Engineering, Lecture Notes in Mechanical Engineering, 11, https://doi.org/10.1007/978-981-16-2086-7_11.
42. Prisha K. Asher, K. Deviparameswari, B. Feonsa Antonitta, S. Meenakshi, R. Vijayanandh, and M. Senthil Kumar, (2021) "Comparative Aerodynamic Performance Analysis on Modified UAV's Propeller by Using CFD", Smart Innovation, Systems and Technologies, 78, 1-12, DOI: 10.1007/978-981-16-0119-4_78.
43. Kumar, M. S., Krishnan, A. S., & Vijayanandh, R, (2021) "Computational, Experimental and Surface Characterisation of Glass Fibre Reinforced Plastic Composite for Wind Turbine Blade Application", Journal of Environmental Protection and Ecology, Vol. 22, Issue 2, pp. 602–616.
44. Ramesh M, Vijayanandh R, Raj Kumar G, Vijayakumar Mathaiyan, P Jagadeeshwaran, Senthil Kumar M, (2021) "Comparative Structural Analysis of Various Composite Materials based Unmanned Aerial Vehicle's Propeller by using Advanced Methodologies", IOP Conf. Series: Materials Science and Engineering, 1017, 012032, 1-10, DOI:10.1088/1757-899X/1017/1/012032.
45. Aswin Kumar. V, Sivaguru. M, Rohini Janaki. B, Sumanth Eswar. K. S, Kiran. P, Vijayanandh R, (2021) "Structural Optimization of Frame of the Multi-Rotor Unmanned Aerial Vehicle through Computational Structural Analysis", IOP Journal of Physics: Conference Series, 1849 012004, https://doi.org/10.1088/1742-6596/1849/1/012004.
46. M. Ramesh, M. Senthil Kumar, R. Vijayanandh, K. Sundararaj, S. Balaji, and S. Bhagavathiyappan, (2021) "Design Optimization of Vertical-Axis Wind Turbine Based on High Power Extraction by Using Computational Fluid Dynamics (CFD)", Innovations in Sus-

tainable Energy and Technology, Advances in Sustainability Science and Technology, 24, 265 - 281, https://doi.org/10.1007/978-981-16-1119-3_24.

4

Design of Vertical Axis Wind Turbine in Recent Years – A Short Review

Vijayakumar Mathaiyan, Vijayanandh Raja, Srinivasamoorthy S, Dong Wong Jung, Senthilkumar M and Sivaranjani Sivalingam

Department of Mechanical Engineering, Jeju National University, South Korea
Department of Aeronautical Engineering, Kumaraguru College of Technology, India

Abstract

As the available models cannot produce the efficiency or power using the horizontal axis wind turbine (HAWT) design, thus, the vertical axis wind turbine (VAWT) design has been reviewed in recent works for its aesthetic value and efficiency. This review is a useful guide to modify the available design for any intended purpose or provide the futuristic design which can be efficient in power generation. Besides, the overview of recent research studies in the field of wind turbine technology is covered in this chapter.

4.1 Introduction

The solar energy from nuclear fusion creates the wind energy. Wind moves because of the atmospheric pressure gradient created by uneven solar heating, Coriolis effect of earth rotation and geographical conditions [1]. The concept of harnessing the energy from wind is one of the most promising technologies to replace the available methodology which pollutes the earth. Recent climatic changes and other adverse effects on the health of human and other living creatures explicitly or implicitly bear testimony to this. With the increasing population and power requirement, the people are not only concerned about the power generation but also the ways to achieve it. We believe that if wind turbines were attractive enough, it would be easy for the employment of such machines and reduce pollution. It is notable that noise from these devices should also be as minimal as possible. Even the governments are encouraging the implementation of the renewable energy sources [2]. The blade thickness and blade loads in the rotor are the major sources of aerodynamic noise. Unsteady flow of VAWT reduces the generated noise, and the maximum power coefficients reaches when $\lambda = 0.5$ [3]. The noise from the wind turbine is a major problem, and its effect on the humans was reported by Amelia et al. Even the tonal components at low frequencies create the discomfort, and the % of the affected people is higher in the case of wind turbines than other sources of noise pollution [4].

Vertical axis wind turbines (VAWTs) are non-responsive to the wind direction due to their omni-directionality. Less moving parts and simple mechanical design are the other advantages. The total energy consumption for the operation and maintenance, stable structure and scalability are the advantages of VAWTs over HAWTs. Difference in blade's circulation along the rotational axis of turbine leads to the energy conversion in VAWTs [5]. Though the power generations from HAWTs is superior than in VAWTs, however, HAWTs have a robust unalterable design. VAWTs are of primary interest only because of their flexible design. In this review, we have provided the VAWT design calculation to meet the power requirement, while providing a view on the available attractive designs. Some of the commonly available VAWTs are Darrieus, Savonius, H-Darrieus-Rotor, helical Darrieus turbine, etc. The authors believe that the helical Darrieus turbine has an attractive design, thus, its design is reviewed here. Further, VAWTs are much safer than HAWTs, if there is any catastrophic event.

This review covers the major design factors and parameters to be considered. We have provided commonly followed design process including the aerodynamic design of blade and effect of low Reynolds number. It is notable that an efficient blade design provides an effective power generation. The lithium batteries can be used as power storage units in renewable energy technologies [6]. Blade element momentum theory, BEM theory application, Navier-Stokes solutions and hybrid methods are the available aerodynamic modelling methods for wind turbines [7].

4.2 Commonly Available VAWTs

Most VAWTs usually have the blade arrangement to create a mechanical energy. Power generation system is directly engaged with the arrangements to generate electricity. A few VAWTs are discussed below.

4.2.1 Vortex Induced Vibrations Wind Turbine

Williamson et al. reviewed work on the vortexes around the circular cylinder and provided more information about the possible vortexes in a vortex induced vibrations wind turbine [8]. The typical arrangements in the VIVWT from Sassi et al. [9] are illustrated in Figure 4.1.

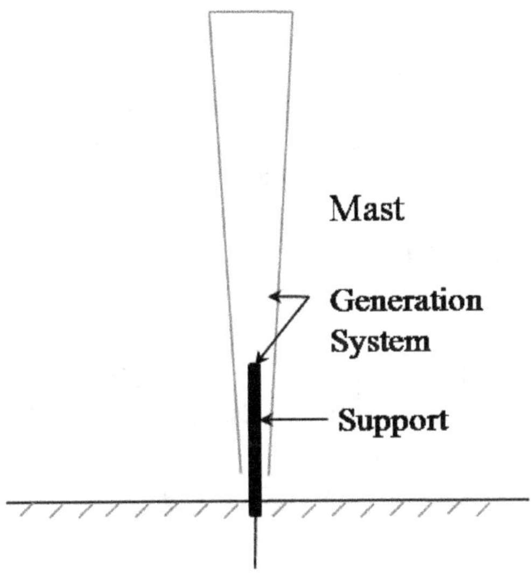

Figure 4.1 Typical arrangement of VIVWT.

The mast, electricity generation system and support system are the components in this bladeless wind turbine. The phenomenon of vorticity creates the mechanical movement of the mast. The mechanical energy is converted into electrical energy using the electricity generation system. A stator is placed on to the supporting system which keeps it stable on the ground, and the rotor is attached to the mast. The arrangement is proved to be efficient than the other available wind turbines. More information about this turbine has also been reported by Villarreal [10].

4.2.2 Tornado Type Wind Turbine

Tornado type wind turbine (TTWT) uses the phenomenon of tornado to generate electricity. As reported by Hsu et al. [11], the setup of the turbine includes the intake chamber, wind chamber, turbine arrangement and vent, as shown in Figure 4.2. The air flows through the intake chamber, and the wind chamber arrangement creates the tornado or vortex, which then hits the turbine. The mechanical energy of the turbine is converted into electrical energy. Yen classified the vortex or tornado as three rotating flow regimes [12]. The viscous effect dominated first flow regime as the tangential velocity increases with radial distance. In the second flow regime, the flow is laminar with almost negligible viscous effects. The circulation for the laminar second flow regime is given by $\Gamma = VRe$, where Re is the Reynolds number, and V is the velocity. Third flow regime has the turbulence with eddies in it. As the reverse flow or ground effect is not desirable, the vent design is also an important design parameter. It is notable that the pathway of air in the wind chamber is affected by the artificial tornado generation systems.

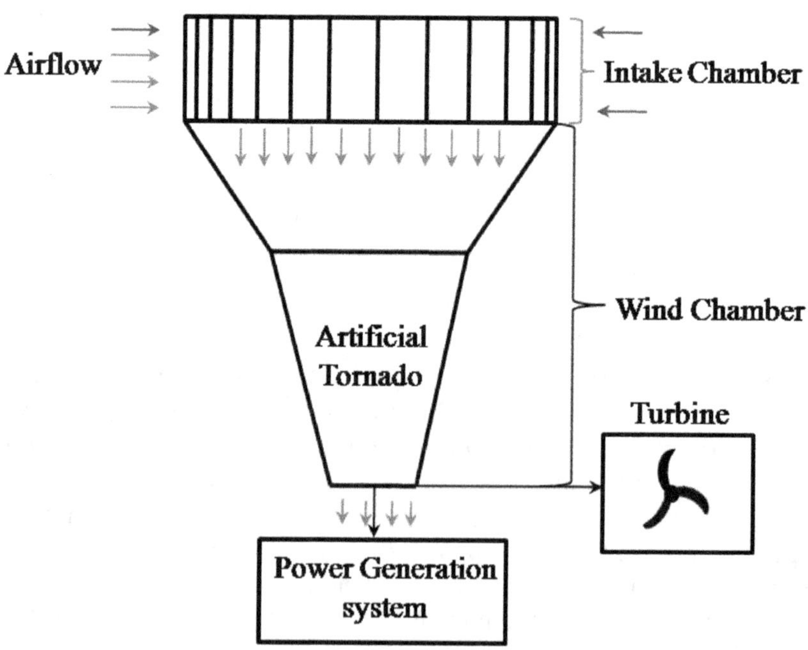

Figure 4.2 Illustration of TTWT.

4.2.3 Darrieus Type Wind Turbine

Tjiu et al. reviewed the Darrieus vertical axis turbines. The Darrieus type wind turbine is classified into curved blades and straight blades type [13]. Guy-wired phi rotor, variable-geometry, variable-pitch, delta rotor, diamond rotor and V/Y rotor are some of the turbines discussed in the review. The small-sized turbines have been researched to reduce the cost of energy, which has resulted in an armless VAWT with looped blades called butterfly wind turbine (BWT). In other words, butterfly type wind turbine is a type of Darrieus type wind turbine with looped blades. Low aerodynamic resistance and blade tip loss is possible in BWT, owing to its armless rotor construction and an additional feature. Over speed control system (OCS) is used to control the speed of BWT under centrifugal force acting on the rotor. OCS is recommended as a prerequisite to reduce the cost of small VAWTs and increase safety. The blade rotates about the blade axis, which is connected to OCS or generator. OCS controls the electric power value around 3.2 kW, even when functioned at 12 m/s or more [14, 15].

Figure 4.3 shows the typical VAWT. As the wind blows, the force of the wind acts on the blade. The blade movement creates the centrifugal force and make the rotor rotate. In the power generation system, the rotor movement cuts the magnetic field in the stator, resulting in power generation.

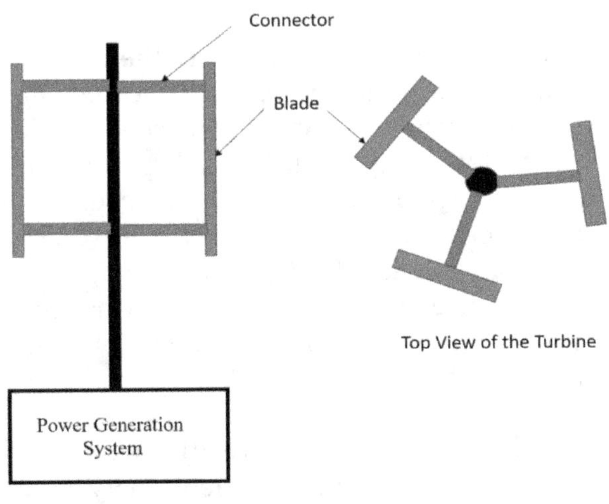

Figure 4.3 Illustration of a typical VAWT.

4.3 Design of VAWT

Designing process involves a series of steps, including the design of VAWT with increased cross-sectional area, selection of materials, airfoil selection, aerodynamic design of blade, number of blades and supporting arrangements, prototype design and computational analysis, etc. Efforts have been made in this work to cover the overall design process of VAWT.

4.3.1 Design Requirements

The major design requirement for VAWT is the maximum power generation, aesthetic design and noise free operation. From Whittlesey et al. [16], the power (*P*) productivity equation for any type of VAWT is given by

$$P = c_p \frac{1}{2} \rho v^3 A \tag{1}$$

where, c_p is the power coefficient, ρ is the density of air, v is the velocity of wind, and A is the cross sectional of VAWT. Equation (1) shows that in order to increase the power generation using design, the cross-sectional area of VAWT should be increased. The cross-sectional area is the radius of the turbine multiplied with the length of blades (Whittlesey et al.). The material employed in making and appropriate placement of VAWT is of major importance.

Arms of a vertical axis wind turbine affect the pressure drag. The blade drag based on pressure is high for the rectangular arm as compared with the aerofoil and cylindrical arms. On the other hand, the pressure based total drag coefficient is high for cylindrical arms, and air foil has low pressure based total drag coefficients [17]. The poor self-starting performance of the lift driven VAWTs can be improved by the double-blade VAWTs. Quadruple-multiple stream tube model (QMS) is used to simulate the performance of DB-VAWT. With a correction factor added to the QMS model, the values are in good compliance with the CFD results [18].

As reported by Schubel et al. [19], the relative wind velocity is given by difference between the wind and blade velocity. Minor losses are noted from the tip, blade aerodynamics from pre- and post-wind impact, driving mechanism, etc. Some of the available VAWT and their predicted efficiency is tabulated below (Schubel et al.)

Table 4.1 VAWT and the peak efficiency from Schubel et al.

VAWT type	Peak efficiency
Savonius rotor	16%
Cup	8%
Darrieus	40%

One should consider the factors such as the material properties, machinability, resistance to the atmospheric effects, safety, availability, cost of material and reliability [20]. The commonly used materials in making wind turbines are carbon fibers, aramid fibers, glass fibers, aluminum, stainless steel and wood. Among the commonly available materials, Maskepatil et al. [21] and Babu et al. [20] claimed that an optimal choice for wind turbine is carbon fibers. The authors considered the factors such as density, mechanical strength, resistance to environmental changes, cost and availability to determine the most suitable material.

Huang et al. [22] reviewed the fabrication and properties of carbon fibers in detail. Carbon fibers are the fibers with at least 92 wt% carbon and have high tensile properties, low density, high conductivity, stability against thermal and chemical stress and resistance to creep. The manufacturing process to fabricate these involves the controlled pyrolysis of the stabilized precursor fibers. The oxidation of the precursor fibers at around 200–400 °C in air stabilizes the material. Subsequently, the resultant material undergoes a process called carbonization. In this process, the non-carbon elements like hydrogen, oxygen, nitrogen, etc., are removed from the stabilized material by treating it at around 1000 °C in an inert atmosphere. Carbonized fibers are then subjected to graphitization at a temperature of about 3000 °C. The Young's modulus of the carbon fibers depends on the carbon content. The carbon fibers are classified based on the Young's modulus and temperature at which these were treated. The atomic structure of carbon fiber and graphite are almost similar. Owing to their properties, the carbon fibers find application in the field of aerospace, automotive industries, and so on.

4.3.2 Blade design

Schaffarczyk et al. [23] discussed the aerodynamics, structural design and manufacturing of the wind turbine blades. The aerodynamics involves the fluid physics on the surface of blade, and the structural design involves the reduction of materials in the blade without compromising the desired properties. The structural integrity of the design is of utmost necessity as the turbines are expected to work for decades. The manufacturing includes the process of machining the blade according to the requirements with minimum cost. Besides the aerodynamic design of the blade, there are other removable accessories like vortex generators and Gurney flaps, employed to enhance the wing aerodynamics. Karimian et al. [24] reported that the 3-part blade design yields better performance than the helical blade with increased tip speed ratio (TSR). Increasing the area of the blades or fins is considered as one of the options to improve aerodynamic performance by Utomo et al. [25]. The size of wind turbines has been increased for generating more energy in the recent years. The optimization of root area of the wind turbine blade is carried using the genetic algorithm, and it is suitable for the complex optimization [26].

4.3.3 Aerodynamics Design

Airfoil Selection

Airfoil selection can be made based on the drag polar or lift drag bucket. A symmetrical airfoil like NACA-0018 is used as it has a low maximum lift coefficient than the asymmetrical airfoil [27]. Castelli et al. studied the blade inclination in Darrieus type wind turbine using NACA 0021 [28]. Selig et al. identified high lift airfoil at low Reynolds number, named as S1223 [29]. The high performance is attained using the concave pressure recovery and aft loading which are the favorable effects. The results exhibit the maximum coefficient of lift $C_{l,\,max}$ = 2.2 at a Re value of 2×10^5. Even the airfoil coordinates are presented in their work. Madasamy et al. reported that the unsymmetrical airfoils can be used in horizontal type wind turbine. These airfoils can also be used for VAWTs [30].

The characteristics of aerofoil determine the power generation of a wind turbine. The flow analyses of aerofoil are used to predict the lift to drag ratio and sound pressure level. The genetic algorithm and Fuzzy Bellman-Zadeh decision making method are the optimization techniques which provide an enhanced lift to drag ratio and a reduction in the sound pressure level [31].

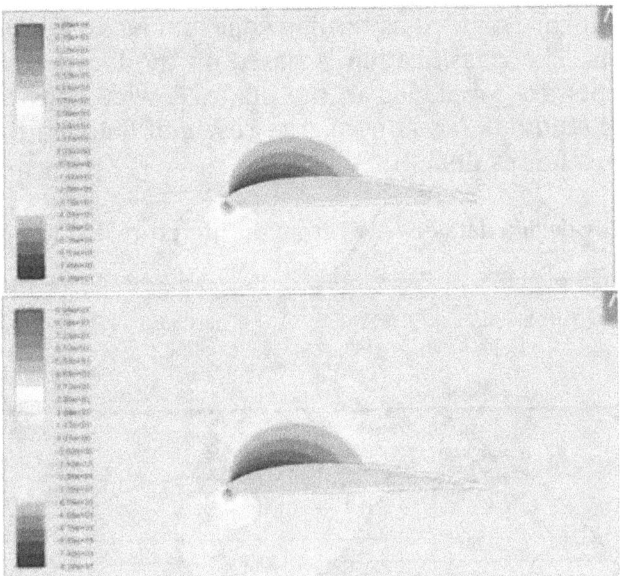

Figure 4.4 The pressure contour over NACA 4412 at $\alpha=4^0$ and $\alpha=6^0$ [30].

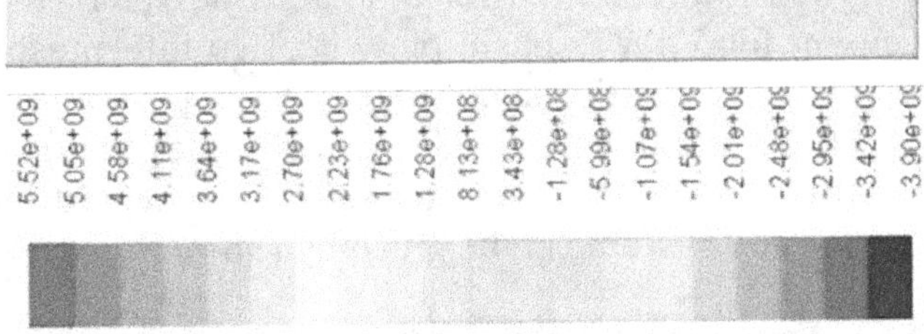

Figure 4.5 The pressure contour for NACA 0018 at velocity of 5 m/s [30].

Methods to Enhance Lift at Low Reynolds Number

Other than designing the airfoil for the desired performance, other techniques like morphing airfoils and additional arrangements such as slats and flaps are employed to attain a high performance. The arrangements are divided as trailing edge and leading-edge high lift mechanisms. The classification is based on the location where the arrangements are employed in the blade. Fowler flaps have been used in the study of Zaccai et al. The research detailed the mechanism with useful results [32].

Table 4.2 The design data of VAWT from the literature [24,33]

Turbine type	Turbine radius [m]	Turbine height [m]	Blade Chord Length [m]	Number of Blades	Airfoil Selection
Straight Blade	0.4	0.8	0.2	3	NACA 0021
3 Part Blade	0.99	1.15	0.3	3	NACA 0021

Some of the mechanisms used in the literature include dropped-hinge, four-bar, link-track and hooked-track. The leading-edge flaps and surface mounted trip can increase lift at low Reynolds number. Employing the leading-edge flaps for the low Reynolds number vehicle is suggested by Jones et al. [34]. Torres et al. showed that the low aspect ratio also leads to a high lift coefficient [35].

4.3.4 Power Transmission Units and Generators

Drive train can be implemented for a smooth power transmission. The arrangement for VAWTs is simple with two intermeshing gear arrangements [36]. The ratios such as gear ratio, transmission ratio and contact ratio should be considered while designing the gear drive for effective power transmission [37]. As reported by Hillier, the gear ratio for any system can be given by [38]

$$Gear\ Ratio = \frac{Number\ of\ teeth\ on\ driven\ wheel}{Number\ of\ teeth\ on\ driving\ wheel}$$

It is observed that most of the VAWT transmission units have fixed gear arrangement. In four wheelers, the number of gear arrangement is variable. The manual or automatic gear transmission is employed to change gears based on the increment in speed. If the automatic gear transmission is utilized, we can operate VAWT in different running conditions.

In case the involute contact ratio is 2 or 3, the design is good to control the noise generated from the spur gears [39]. If the size of the wind turbine is large, the lubrication of the gear drive should be considered.

4.4 VAWT Farms

Wind turbines in large arrays reduce the efficiency of the turbines. Large eddy simulation is the model used to study the wind-turbine array boundary layers [40]. Hezaveh et al. carried out a parametric study on the VAWT farms and revealed that the triangular arrangements are desirable [41]. The results showed that the triangular design makes use of the flow acceleration and has a minimum effect. The parametric analysis of Hezaveh et al. for the VAWT farm is also shown in Figure 4.6. It is concluded that the normalized coefficient of pressure value is high for the cluster staggered (60 degree) and low for the aligned one. The staggered cluster has a C_P^* value higher than the aligned and lower than the cluster staggered (0 degree). C_P^* is the direct measure of the amount of power generated from the wind farms. C_P^* was introduced in their study by Hazaveh et al., and the normalized relation of the quantity per unit power input was given as

$$C_P^* = \frac{2P}{\rho A u^2 U} \tag{2}$$

$$u^2 = \Delta P_{drop} \frac{L_x}{L_z} \tag{3}$$

where, P is the average power per VAWT, ρ is the density of the air, u is the root mean square of the total mean drag, ΔP_{drop} is the total pressure drop in the VAWT farm, U is the averaged streamwise velocity component, A is the farm area, L_x is the length of the farm, and L_z is the height of VAWT.

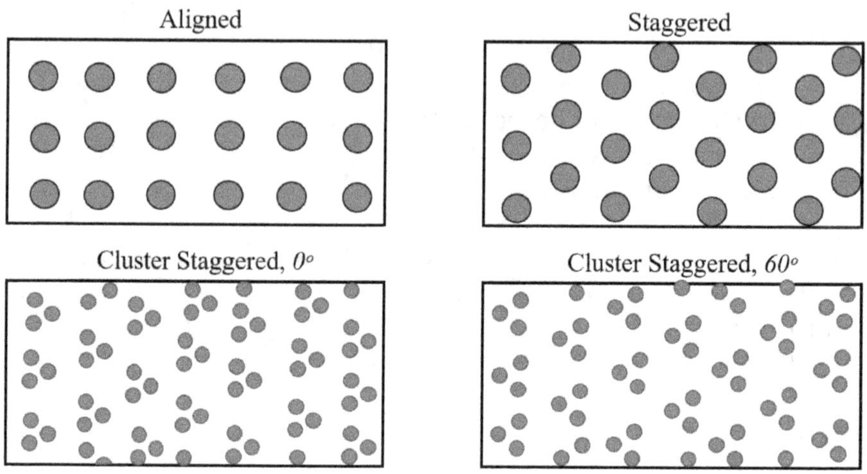

Figure 4.6 The parametric study arrangement of the VAWT wind farm reported by Hezaveh et al.

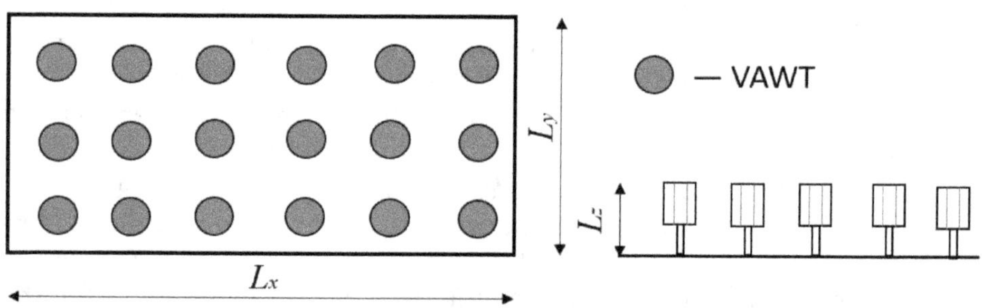

Figure 4.7 The dimensions of the VAWT farm [41].

The wind power forecasting is replaced with a technique called advance neural networks (ANN) [42].

$$x_i(t+1) = a_i x_i(t) + b_i \boldsymbol{\Sigma} w_{i,j}(t) y_j(t)$$

where, *a* and *b* are the constants, *x* is the state of the respective neuron, w is the synaptic weight for the jth input of ith neuron, and *y* is an external input or state of neuron inputted from the sigmoidal function. The wind park power profile, an important factor used to measure the power performance, is the ratio of the power from total number of wind turbines at a certain time to the power from the number of wind turbines operated at the time. The prediction of this quantity is difficult because of the external factors, and the measurement is made based on the data acquired from the farm or park. The test is carried with the three-basic configurations, viz. the 12 outputs from a network for every time step, 1 output network with 12-time steps forecasted iteratively and error optimized network of the 2nd configuration. Though HAWT farms are widely used worldwide, however, the VAWT farms should be considered for their efficiency. For instance, the HAWT farms in a few areas in India generate considerable amount of power. The top 5 states contributing to the power generation with its capacity is tabulated in Table 4.3 [43]. Further, Korea Electric Power Corporation (KEPC) is carrying out notable works to generate green energy in South Korea. The countries like Japan are prone to the natural disasters, thus, it is important to keep these external factors in the design for a reliable structure.

Table 4.3 Top contributors of wind energy in India

State	Capacity in M
Gujarat	84431
Rajasthan	18770
Maharashtra	45394
Tamil Nadu	33799
Madhya Pradesh	10483

4.5 Numerical Simulation for the VAWT Design

Commercial software packages like Ansys, Hyper Works, Abaqus, etc., are used to visualize the fluid dynamics of VAWTs. The visualization provides information about the changes which may be required for better aerodynamic efficiency. Though the theoretical calculations equip us with data, the commercial software provide better insights to proceed further in the designing process. The pisoFoam Turbine solver is an apt choice for the analysis when the horizontal gradient pressure and Coriolis force in momentum equation are negligible. ALM (actuator line model) captures the flow simulation around the blades in a wind farm of 12 wind turbines [44].

The usage of blade element theory in the aerodynamic load of an offshore floating wind turbine may not be accurate because of the complex phenomena such as ground effects, variations in wind shear, unsteady yaw and aerodynamics, etc. Coding for unsteady vortex lattice method is integrated with the FAST AeroDyn software and has been validated. This method has an acceptable level of accuracy and performance in unsteady yaw condition [45].

Ansys Fluent covers a wide range of fluid problems. From the literature review [46], the best model to analyze the wind turbine is the shear stress transport (SST) kinetic energy k and dissipation ω model. Menter et al. reported that this model can be used for the prediction of adverse pressure gradient flows [47].

$$\frac{\partial(\rho k)}{\partial t} + \frac{\partial(\rho k u_i)}{\partial x_i} = \frac{\partial}{\partial x_j}\left(\Gamma_k \frac{\partial k}{\partial x_j}\right) + G_k - Y_k + S_k \tag{4}$$

$$\frac{\partial(\rho \omega)}{\partial t} + \frac{\partial(\rho \omega u_i)}{\partial x_i} = \frac{\partial}{\partial x_j}\left(\Gamma_\omega \frac{\partial \omega}{\partial x_j}\right) + G_\omega - Y_\omega + S_\omega \tag{5}$$

where, G_k is the generation of the turbulence kinetic energy due to mean velocity gradient, G_ω is the generation of specific rate of dissipation, S is the source terms, Y is the dissipation due to turbulence, u is the velocity term, and Γ is the effective diffusivities.

$$\Gamma_k = \mu + \frac{\mu_t}{\sigma_k} \tag{6}$$

$$\Gamma_\omega = \mu + \frac{\mu_t}{\sigma_\omega} \tag{7}$$

The turbulent viscosity μ_t is calculated using the following equation.

$$\mu_t = \frac{\rho k}{\omega} \frac{1}{\max\left[\frac{1}{\alpha^*}, \frac{SF_2}{a_{1\omega}}\right]} \tag{8}$$

where, S is the strain rate magnitude.

$$\sigma_k = \frac{1}{F_1/\sigma_{k,1} + (1 - F_1)/\sigma_{k,2}} \tag{9}$$

$$\sigma_\omega = \frac{1}{F_1/\sigma_{\omega,1} + (1 - F_1)/\sigma_{\omega,2}} \tag{10}$$

The blending functions F_1 and F_2 are provided for a distance y to the next surface as below.

$$F_1 = tanh(\phi_1{}^4) \tag{11}$$

$$\phi_1 = min\left[max\left(\frac{\sqrt{k}}{0.09\omega y}, \frac{500\mu}{\rho y^2\omega}\right), \frac{4\rho k}{\sigma_{\omega,2} D_\omega^+ y^2}\right] \tag{12}$$

$$D_\omega^+ = max\left[2\rho\frac{1}{\sigma_{\omega,2}}\frac{1}{\omega}\frac{\partial k}{\partial x_j}\frac{\partial \omega}{\partial x_j}, 10^{-10}\right] \tag{13}$$

$$F_2 = tanh(\phi_2{}^4) \tag{14}$$

$$\phi_2 = \left[max\left(\frac{2\sqrt{k}}{0.09\omega y}, \frac{500\mu}{\rho y^2\omega}\right)\right] \tag{15}$$

where, D_ω^+ is the cross-diffusion term.

The production of turbulence kinetic energy is $G_k^* = \min\left(G_k, 10\rho\beta^* k\omega\right)$

$$G_\omega = \frac{\alpha}{v_t} G_k^* \tag{16}$$

The constant $a_\infty = F_1 a_{\infty,1} + (1 - F_1) a_{\infty,2}$

$$a_{\infty,1} = \frac{\beta_{i,1}}{\beta_\infty^*} - \frac{k^2}{\sigma_{\omega,1}\sqrt{\beta_\infty^*}} \tag{17}$$

$$a_{\infty,1} = \frac{\beta_{i,2}}{\beta_\infty^*} - \frac{k^2}{\sigma_{w,2}\sqrt{\beta_\infty^*}} \tag{18}$$

Dissipation of ω,

$$Y_k = \rho\beta^* k\omega \tag{19}$$

$$Y_k = \rho\beta^*\omega^2 \tag{20}$$

The value of β_i is calculated using the following relation.

$$\beta_i = F_1\beta_{i,1} + (1 - F_1)\beta_{i,2} \tag{21}$$

The cross-diffusion modification D_ω,

$$D_\omega = 2(1 - F_1)\rho\sigma_{\omega,2}\frac{1}{\omega}\frac{\partial k}{\partial x_j}\frac{\partial \omega}{\partial x_j} \tag{22}$$

The value of the model constant is given below,

$$\sigma_{k,1} = 1.176, \sigma_{\omega,1} = 2.0, \sigma_{k,2} = 1.0, \sigma_{\omega,2} = 1.168$$

$$a_1 = 0.31, \beta_{i,1} = 0.075, \beta_{i,2} = 0.0828$$

The other constants in the above equations remain the same for the model.

User defined functions (UDF) in Fluent allow us to analyze the rotatory component using the C program. The function usually requires the boundary conditions, which must be imposed on to the desired moving components. Figures 4.8 and 4.9 show the computational domain used for such analysis. Using UDF in Fluent, the authors imposed the angular velocity to the airfoils, and the analysis was carried out subsequently. When the analysis is made with moving object in the computational domain, dynamic mesh is used. After running the computational fluid dynamics, Ansys allows to import the values for the structural analysis. The technique is called as fluid structure interaction (FSI).

The free rotation of the wind turbine blades completely depends upon the rigid support and its behavior. Owing to this reason, only lightweight cum high strength-based materials have been proposed for the construction of the modern wind turbines. The ultimate motto of the wind turbine is to extract a high amount of wind energy from the free stream velocity of the fluid through aerodynamic rotors. With the help of the aerodynamically efficient design profiles of the rotors, the induced aerodynamic forces and thereby rotation of the wind turbine rotors are increased at an acceptable rate. By means of the implementation of the lightweight cum high strength materials in wind turbines, the induced aerodynamic forces are successfully converted into real-time movement of the rotors. Therefore, the investigation about the suitable lightweight material for the use of real-time applications and fatigue life estimations are much needed in the wind energy sector. The working environment of the wind turbine is very complex, which is the cumulative impact of the aero- and rotodynamic effect, so a normal approach does not have the capacity to capture the external aerodynamic load for structural simulation. Thus, an advanced approach can have the capability to overcome the aforesaid critical environment-based problems, namely FSI based coupling computation. In FSI, the broad environments are coupled effectively in order to attain the structural outcomes of the given wind turbine rotors, thereby providing a high performance of rotors with respect to equivalent stress, fatigue life, strain energy, etc. Through the help of an advanced coupling scheme, the complete aerodynamic load is transferred from the fluid environment to the structural environment, which enhances the reliability of the structural outcomes. Based on the way of the coupling scheme, two major classifications are available in the advanced computations, which are one way coupling and two-way coupling. One way coupling facility based FSI is mostly available in all kinds of FEA and CFD solvers, whereas the two-way coupling facility based FSI is suitable and adaptable for the advanced computational solvers. As the two-way coupling needs to be computed by means of a separate unique tool, it is named as the system coupling. Comparatively, two way coupling approach is the best pathway to provide trustworthiness structural outcomes on wind turbines. Two way coupling approach is based on parallel computation. Thus, unless otherwise good boundary conditions are used, the attainment of results is very tough. On the other hand, good boundary conditions allow the production of actual results, thus, two-way coupling is more preferable to be implemented in the wind turbine problems.

The input inlet velocity Vi is given as the inlet boundary condition and boundary chosen for outlet is pressure far field. A constant angular velocity ω is imposed on the blade arrangements. The number of elements used in the computational domain is 256014, and the time step of 0.0016 s is simulated with the 10^4 - time steps. The converged solution has the input velocity of 7 ± 1 m/s, and the blades are set to rotate at a constant angular velocity of 10 rad/s.

The velocity contour, obtained by solving the above computational grid, shows that there is a considerable pressure variation on the blades because of the provided inlet velocity. This shows that a notable change in the rotational speed of the turbine blades results in the generation of more power. The input velocity is chosen as 7 m/s, as it is the average wind speed in this work.

The computational domain is the 3 turbine blade arrangement in 2D, excluding the mounting equipment. The result from the preliminary research in 2D is in a good compliance with the results from other research articles. Owing to the computational cost, the authors have performed the research on computational domain with UDF. UDF imposes the angular velocity on the blade using the following equation.

$$\omega = \frac{V\lambda}{R}$$

where, V is the freestream velocity, λ is the tip speed ratio of the blade, and R is the radius of the blade. In Fluent software package, either DEFINE_ZONE_MOTION or DEFINE_TRANSIENT_PROFILE have been used. Both UDFs allow the rotational rate of the domain with the angular velocity and time t. UDF also allows us to input unsteady pressure, which is the real time case for the wind turbine. There is always an unsteady airflow flow in the earth atmosphere. Though there are a plenty of sophisticated commercial packages available to perform the computational analysis, however, one should be careful while validating the results. The common errors originate from the mesh quality, dirty geometry, model used for analysis, etc. The experimental and/or analytical validation is always recommended even if one is convinced with the computational results. An example of UDF coding for the unsteady pressure input is shown below. The information such as the profile definition and boundary to face is provided in the program.

```
*****************************************************************
**********

                        UDF CODING
      (c programming language integrated with Fluent)
*****************************************************************
**********
```

```
#include "udf.h"
  DEFINE_PROFILE (Unsteady_Pressure, thread, position)
  {
     face_t f;
     real t = CURRENT_TIME;
     begin_f_loop(f, thread)
     {
        F_PROFILE (f, thread, position) = 101325.0 + 7.0*sin(10.0*t);
     }
     end_f_loop(f, thread)
   }
*******************************************************************
*********
```

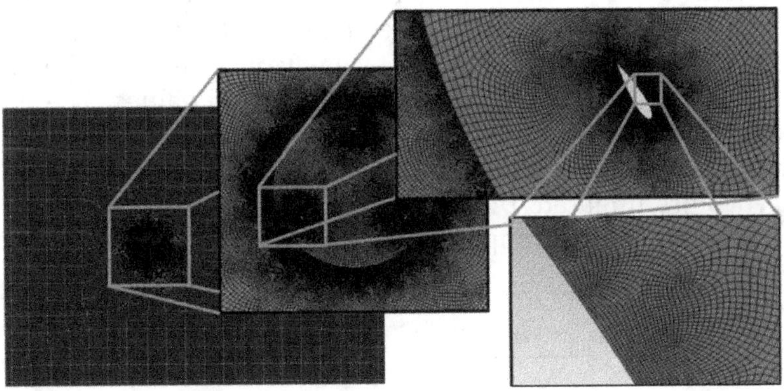

Figure 4.8 The computational domain as reported by Madasamy et al. [30].

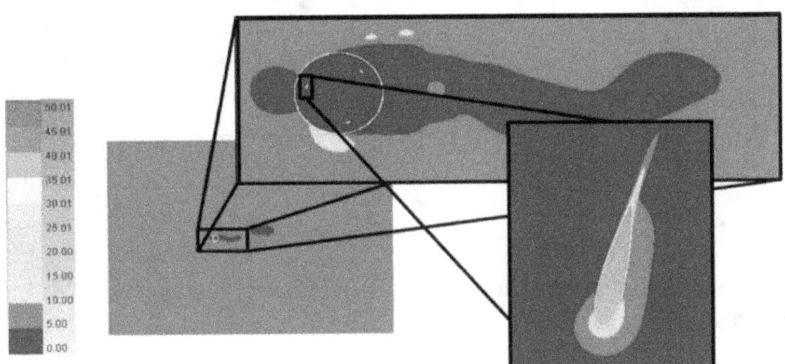

Figure 4.9 The velocity (V) contour of the turbine blades rotating at an angular velocity of 10 rad/s where input velocity is 7 ± 1 m/s [30].

4.6 Computational Method

Pressure based Segregated Solver

The solution algorithm employed in pressure based segregated solver solves the governing equations sequentially. A solution loop is carried out iteratively to attain a converged numerical solution. The nonlinear and coupled governing equations are solved one after another. The solution procedure is given below.

Solution Procedure

When a two-step initialization sequence is executed outside the solution iteration loop, the solution process commences. The inputs from the user interface initialize equations after the sequence has begun. PROFILE and INIT UDFs are called consecutively. The initialization overwrites the previously set initialized values. The execution of PROFILE UDFs and Adjust UDFs is carried out when the loop begins. The governing equations are solved segregatively, and the properties are updated. Depending upon the solution convergence, the continuation of the loop is determined [46].

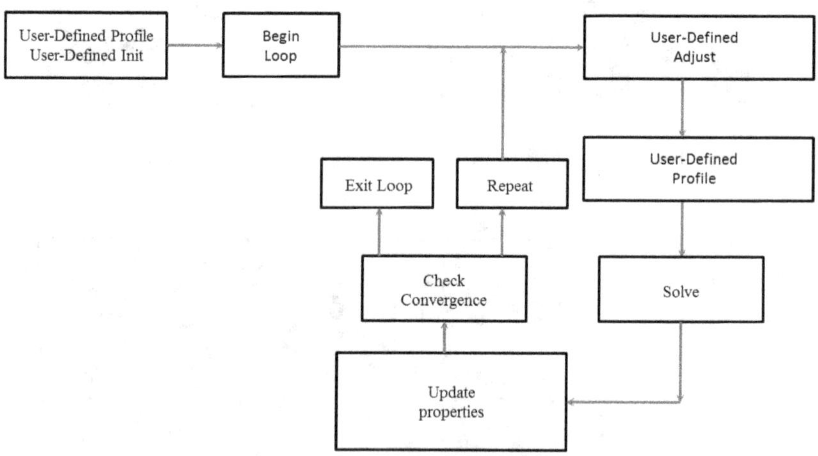

Figure 4.10 The flow chart of solution procedure based on the SIMPLE scheme [46].

Niculescu et al. showed that the existing turbulence model is not suitable for the Turby Darrieus wind turbine and suggested that proper orthogonal decomposition (POD) can explain the different projection in the decomposition process [48]. Consequently, the analysis is carried to the finite dimensional domain by restricting the approach to the separation of variables in space and time [49].

Experimental Methods for Validation

Smoke visualization, hotwire measurement, and thrust force measurement are the techniques used to experimentally study VAWTs in skewed flow condition. For the flow visualization, the smoke is injected in the upwind of rotor, and a stroboscopic light is used to analyze the skewed flow. A hotwire is employed to determine the velocity. The thrust is measured by balancing the VAWT model, and for the skewed model, the measurements are based on the velocity in and perpendicular to the flow [50].

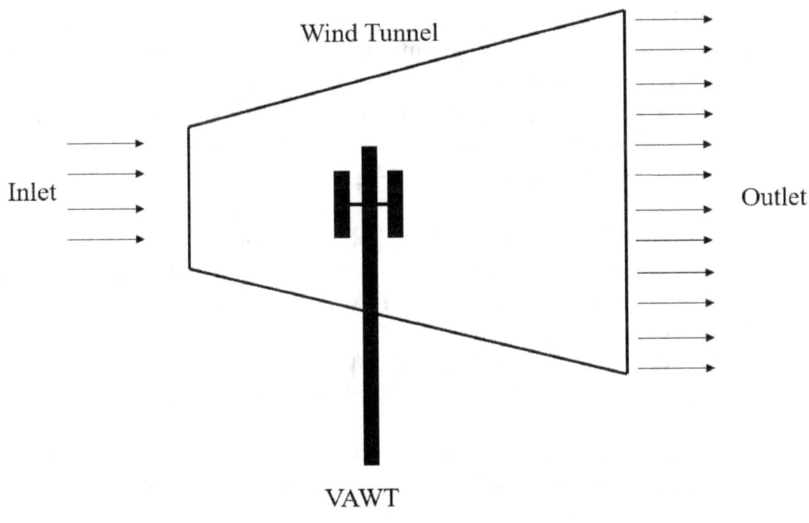

Figure 4.11 The typical wind tunnel arrangement for the validation of results.

In a typical wind turbine, the inlet, wind tunnel and outlet from the main components. The wind is forced out from the inlet using a fan in the outlet. VAWT to be tested is placed inside the wind tunnel for the possible flow visualization or other measurements. The arrangements inside the wind tunnel usually include contraction, test and diffusion parts. There are various types of wind tunnels based on the speed of fluid flow. If the flow is steady and incompressible, neglecting the viscosity, the speed of the fluid V is given by,

$$V = \sqrt{\frac{2\Delta p}{\rho}} \tag{23}$$

where, Δp is the pressure difference between the total and static pressure, and ρ is the fluid density.

References

1. Tong, Wei, (2010) Wind Power Generation and Wind Turbine Design, WIT Press.
2. Weiss, Jurgen, Tsuchida, Bruce T., (2017) Integration into National Grids, Wind Energy Engineering, A Handbook for Onshore and Offshore Wind Turbines, pp. no. 419 – 436.
3. A. Dumitrache et al., (2015) Influence of Unsteady Flow on the Aerodynamics of Vertical Axis Wind Turbines, AIP Conference Proceedings, pp. no. 500007 – 1 – 4.
4. Trematerra, Amelia et al., (2017) Wind Turbines Acoustic Measurements, AIP Conf. Proceedings, pp. no. 020001 – 1 – 7.
5. Victor, Mendoza et al., (2017) Wake Flow Simulation of a Vertical Axis Wind Turbine Under the Influence of Wind Shear, J. Phys.: Conf. Ser., pp. no. 012031 – 1 – 10.
6. Meng, Huixing, Li, Yan-Fu, (2019) A Review on Prognostics on Health Management (PHM) Methods of Lithium-ion Batteries, Renewable and Sustainable Energy Reviews, pp. no. 1 – 12.
7. Das, P.M. Mohan, Amano, R.S., (2014) Basic Theory for Wind Turbine Blade Aerodynamics, Aerodynamics of Wind Turbines, WIT press, Vol. 81, pp. no. 11 – 23.
8. Williamson, C.H.K., and Govardhan, R., (2004) Vortex Induced Vibrations, Annual Review of Fluid Mechanics, 36(1), pp. no. 413 – 455.
9. Sassi, Paolo, Freiria, Jorge, et al., (2020) Simulation of Vorticity Wind Turbines, Heliyon, 6, pp. no. 1 – 12.
10. David Jesus Yanez Villarreal, (2016) Vortex Resonance Wind Turbine, US patents, pp. no. 1 – 9.
11. Hsu, Cheng-Ting, (1984) Tornado Type Wind Turbines, US patent, pp. no. 1 – 4.
12. Yen, James T., et al., (1978) Tornado-Type Wind Turbine, US Patent, pp. no. 1 - 8.

13. Tjiu, Willy, et al., (2015) Darrieus Vertical Axis Wind Turbine for Power Generation I: Assessment of Darrieus VAWT Configurations, Renewable Energy, 75, pp.no. 50 – 67.
14. Hara, Yutaka, et al., (2018) Development of a Butterfly Wind Turbine with Mechanical Over-Speed Control System, MDPI publishing, Vol. 2, pp. no. 1 – 25.
15. Hara. Y, et al., (2017) Numerical Simulation on Fluid Forces and Structure of Triangular-Blade Butterfly Wind Turbine, ICJWSF, October, Ohio, USA.
16. Whittlesey, Robert, (2017) Vertical Axis Wind Turbines: Farm and Turbine Design, A Handbook for Onshore and Offshore Wind Turbines, Wind Energy Engineering, pp. no. 185 – 202.
17. Hara, Yutaka, Horita, Naoki, et al., (2019) Numerical Analysis of Effects of Arms with Different Cross-sections on Straight-Bladed Vertical Axis Wind Turbine, Energies, MDPI publication, pp. no. 1 – 24.
18. Hara, Yutaka et al., (2014) Predicting Double-Blade Vertical Axis Wind Turbine Performance by a Quadruple-Multiple Stream tube Model, International Journal of Fluid Machinery and Systems.
19. Schubel, Peter J., Crossley, Richard J., (2012) Wind Turbine Blade Design, Energies, MDPI Journals, 5, 3425 - 3449.
20. Babu, Suresh. K., Subba Raju, N.V., et al., (2006) The Material Selection for Typical Wind Turbine Blades a MADM approach and Analysis of Blades, MCDM, pp. no. 1 – 12.
21. Maskepatil, Lahu P., Gandigude, A.U., Kale, S.A., (2014) Selection of Material for Wind Turbine Blade by Analytic Hierarchy Process (AHP) Method, Applied Mechanics and Materials, Trans Tech Publications, pp. no. 145 – 150.
22. Huang, Xiaosong, (2009) Fabrication and Properties of Carbon Fibers, Materials, MDPI publishing, pp. no. 2369 – 2403.
23. Schaffarczyk, A.P., (2014) Introduction to Wind Turbine, Green Energy and Technology, Springer Publications.
24. Kariamian, S.M.H., Abdolahifar, Abolfazl, (2019) Performance Investigation of a New Darrieus Vertical Axis Wind Turbine, Energy, pp. no. 1 – 48.
25. Utomo, Ilham Satrio et al., (2018) Experimental Studies of Savonius Wind Turbines with Variations Sizes and Fin numbers towards Performance, AIP Conference Proceedings, pp. no. 030041 – 1 – 7.
26. Herrmann, Jordis et al., (2019) Multi-objective Optimization of a Thick Blade Root Airfoil to Improve the Energy Production of Large Wind Turbines, Journal of Renewable and Sustainable Energy, 043304 – 1 – 25.
27. Shahariar, Hasan G.M., Hasan, Mohammad Rashedul, (2014) Design and Construction of a Vertical Axis Wind Turbine, The 9th International Forum on Strategic Technology (IFOST), pp. no. 326 – 329.

28. Castelli, Marco Raciti, et al., (2012) Effect of Blade Inclination Angle on a Darrieus Wind Turbine, Journal of Turbomachinery, ASME, pp. no. 031016 – 1 – 10.
29. Selig, Michael S., and Guglielmo, James J., (1997) High-Lift Low Reynolds Number Airfoil Design, Journal of Aircraft, Vol. 34, No. 1, pp. no. 72 – 79.
30. Madasamy, Senthil Kumar, et al., (2021) Investigation of NACA airfoils for Aerodynamic Performance of Wind Turbines, pp. no. 1 – 12.
31. Sanaye, Sepehr et al., (2014) Multi-objective Optimization of Airfoil Shape for Efficiency Improvement and Noise Reduction in Small Wind Turbines, Journal of Renewable and Sustainable Energy, pp. no. 053105 – 1 – 21.
32. Zaccai, David, Bertels, Francesco, Vos, Roelof, (2017) Design Methodology for Trailing-edge High Lift Mechanisms, CEAS Aeronaut J., pp. no. 521-534.
33. Elkhoury, M., Kiwata, T., Aoun, E., (2015) Experimental and Numerical Investigation of a Three-dimensional Vertical-axis Wind Turbine with Variable-pitch, Journal of Wind Engineering and Industrial Aerodynamics, pp. no. 111 – 123.
34. Jones, A. R., et al., (2008) Low Reynolds Number Aerodynamics of Leading-Edge Flaps, Journal of Aircraft, Vol. 45, No. 1, pp. no. 342 – 345.
35. Torres, Gabriel E., and Mueller, Thomas J., (2004) Low-Aspect-Ratio Wing Aerodynamics at Low Reynolds Numbers, AIAA Journal, Vol. 42, No. 5, pp. no. 865 - 873.
36. Govind, Bala, (2017) Increasing the Operational Capability of a Horizontal Axis Wind Turbine by its Integration with a Vertical Axis Wind Turbine, Applied Energy, pp. no. 479 – 494.
37. Jelaska, Damir, Gears and Gear Drive, A John Wiley & Sons publication, 2012.
38. Hillier, V.A.W., (2012) Hillier's Fundamentals of Moto Vehicle Technology 6th Edition, Nelson Thornes Ltd publication.
39. Kahraman, A., Blankenship, G. W., (1999) Effect of Involute Contact Ratio on Spur Gear Dynamics, Transactions of the ASME, pp. no. 112 – 118.
40. Calaf, Marc, et al., (2010) Large Eddy Simulation Study of Fully Developed Wind-Turbine Array Boundary Layers, Physics of Fluids, pp. no. 015110 – 1 – 17.
41. Hezaveh, Sayed Hossein, et al., (2018) Increasing the Power Production of Vertical-Axis Wind-Turbine Farms Using Synergistic Clustering, Boundary-Layer Meteorology, Springer, pp. no. 1 – 22.
42. Kariniotakis, G. N., et al., (1996) Wind Power Forecasting Using Advanced Neural Networks Models, IEEE Transactions on Energy Conversion, Vol. 11, Issue. 4, pp. no. 762 – 767.

43. https://www.mnre.gov.in/wind/current-status/.
44. Strijhak, S V, et al., (2019) 'Simulation of Turbulent Wakes in Model Wind Farm with Arbitrary Location for Wind Turbines', Journal of Physics: Conference Series, 1382, pp. no. 012043 – 1 – 7.
45. Jeon, Minu et al., (2014) Unsteady Vortex Lattice Method Coupled with a Linear Aeroelastic model for Horizontal Axis Wind Turbine, Journal of Renewable and Sustainable Energy, 042006 – 1 – 7.
46. Fluent Manual, Ansys 12, 2009, Ansys Inc, PA.
47. Menter, F. R., Two-Equation Eddy-Viscosity Turbulence Models for Engineering Applications, AIAA Journal, Vol. 32, No. 8, 1994, pp. no. 1598 – 1605.
48. Niculescu, M. L., et al., CFD analysis of a Darrieus Wind Turbine, AIP Conference Proceedings, 1863, 2017, pp. no. 420003-1 – 4.
49. Weiss, Julien, A Tutorial on the Proper Orthogonal Decomposition, AIAA conference publication, 2019, pp. no. 1 – 21.
50. Ferreira, Carlos, et al., Wind Tunnel Hotwire Measurements, Flow Visualization and Thrust Measurement of a VAWT in Skew, 44th AIAA Aerospace Sciences Meeting and Exhibit, 2006, pp. no. 1 – 12.

5

A Primer on Renewable Energy Financing

Labanya Prakash Jena[1] and Shreyans Jain[2]

Climate Policy Initiative, India Private Limited, Saket, New Delhi, India

5.1 Introduction

The global energy order is undergoing a profound transformation. Governments and corporates worldwide are moving away from carbon-emitting fossil fuels towards carbon-neutral renewable energy (Frankfort School, 2017) [1]. According to the International Renewable Energy Agency (IRENA), the share of renewables would need to increase to 65% of the global energy supply by 2030 to meet the Paris Agreement on limiting temperature below 2 degrees above the pre-industrial level by 2050. One of the key challenges to achieving the globally agreed climate objectives is the mobilization of adequate capital. The amount of capital flowing to the renewable energy sector, at present, is not enough to meet the target. There is a need to make necessary interventions to accelerate capital flows to the sector, including treating the sector differently from conventional fossil-fuel-based energy.

The financing needs of renewable energy are different from the conventional energy sector. The renewable energy projects are capital-intensive and require large up-front capital, making the sector different from the conventional energy sector. At the same time, the sector is also vulnerable to several risks, and some are unique and new to finance professionals. Appraising the commercial viability of the renewable project and choosing the most efficient financing mechanism is a challenge. Financiers, therefore, need to have a clear understanding of the economics of renewable energy projects to make informed decisions.

This chapter is a primer on renewable energy financing. The first section of the chapter covers the means of renewable financing and the different financing instruments deployed to raise capital for renewable projects. The second section explains the impact of the project financing approach on the project's risk and returns profile and discusses the reasons that have made the project finance a preferred choice of financing. The third section covers the structuring of a renewable energy project using the project finance approach. It is followed by a discussion on the cash flow trapping mechanism in section four. Finally, the fifth section investigates various risks associated with the renewable energy projects at each stage of development and discusses risk mitigation strategies.

5.2 Means of Renewable Energy Financing

Historically, the government, bilateral and multilateral development finance institutions were the primary financiers of renewable energy projects. The sector was not appealing to private financiers from an investment perspective. Due to their limited participation, the renewable energy projects were considered frontier technology investments by investors. However, rapid technological advances in the last five years and proof of performance have helped eliminate several risks associated with the renewable energy sector. Renewable energy investment, at present, is very similar to electric utility-like investments, where risk is much lower than fossil fuel power generating assets. In a typical corporate finance approach, project developers funded their capital expenditure requirements through balance-sheet financing. In this approach, the financial strength and credibility of the parent company played a major role in determining the characteristics of debt and the terms attached to it. The lenders provided funds to the project sponsor against the project's cash flows and the parent company's assets. The repayment profile was drawn against the balance sheet of the corporate. In the event of default, lenders had a full claim on the asset portfolio of the parent company to generate cash flows to service their loans. This arrangement often led to risk contamination and reduced the claim of existing shareholders.

However, in a project finance approach, the renewable energy project becomes a standalone business unit, called a special purpose vehicle (SPV). Structuring a project through this approach helps allocate and mitigate risks and returns by providing bankruptcy remoteness to project assets, contracts and cash flows. Each project is a separate legal entity in this financing structure. A highly debt-driven project finance mechanism lowers the cost of capital, improves credit discipline by placing stringent covenants and enhances the return on the capital invested (Nevitt and Fabozzi, 2000) [2]. The volatility of cash flows associated with the project is low, and there is a definite demand and off-take of the energy generated by way of power purchase agreements.

Low risks have attracted many private financiers to the sector. In addition to traditional financial instruments such as debt and equity, new innovative financial instruments such as green bonds and yield companies (Yieldco) are gaining prominence among investors. The rise of these financial innovations is driven by strong government support. These financial innovations are attracting new classes of investors to the sector, who were earlier reluctant to provide capital through traditional financial instruments. This financial innovation has advanced the sector's liquidity profile in the market and reduced the financing cost. The key financing instruments are discussed as follows:

5.2.1 Debt Financing

The renewable energy generation sector offers greater stability and predictability in cash flow than other sub-sectors in the broad infrastructure sector (Jena and Meattle, 2018) [3]. High predictability in sales volume and pricing through long-term power purchase agreements, low and stable operating expenses, no input cost (zero fuel expenses) and no recurring capital expenditure make renewable energy more reliable than any other sub-sector of infrastructure assets. The stability of the renewable energy sector enables developers to raise a large amount of debt financing. The source of debt financing are:

- **Corporate or Project Debt** - From a lender's perspective, project debt financing is better as they have full control over the project cash flow, thus, monitoring the project is easy. They can tie up management behavior (covenants and restrictions) with a particular risk. However, the lender also faces a high default risk if the project generates less cash than estimates. The lender's capital is not exposed to a particular project in the corporate financing model and returns from other projects can cover any cash shortfall in one project.

- **Loans or Bonds** - Commercial banks and financial institutions are the two major sources of debt financing. Besides, domestic and international development financial institutions provide debt capital cheaply to renewable energy projects in developing countries if they are not commercially feasible for private financiers. As the renewable projects mitigate carbon emission, these financial institutions mandate to finance the projects, which generate positive externalities (reducing carbon emission). For instance, International Finance Corporation (IFC), Asian Development Bank (ADB) and Green Finance Fund (GCF) provide debt financing to renewable energy projects in developing countries like India. In addition, government-backed financial institutions also offer debt capital on favorable terms to these projects as they are committed to increasing renewable energy in their energy fix.

  In bond issuance, corporates have access to a wide investor base that could lead to better credit terms, including decreasing costs and increasing the tenor of funding. Renewable energy projects need long-term debt financing, usually for the asset's useful life of 25 years. Corporates raise funding from commercial banks for a medium-term (5-7 years) as banks hesitate to lend for a long period due to their short liability duration. Hence, corporates face refinancing risk if they borrow capital from commercial banks. Issuance of bonds is an attractive proposition for corporates as it allows them to raise capital for long-term, more than 20 years. Pension funds, insurance companies, credit funds and mutual funds are the primary investors in bonds. Pension and insurance companies usually invest in long-duration bonds to match their long-duration liabilities (Hertrich, 2013) [4]. Renewable energy sector bonds also mitigate the asset-liability mismatch risk of liability-driven investment funds. The long duration of renewable project investment can reduce reinvestment risk from the perspective of investors. However, the corporates should be careful of the rating of the bonds. If the rating of the bond is too low, pension and insurance firms do not invest in these bonds as their respective regulators usually prohibit them from investing in low-rated bonds. The low rating of these bonds would increase the yield of these bonds in the primary market, which would increase the cost of funding for the issuer.

- **Green Bond** - A green bond is a fixed-income instrument for private institutions to raise debt capital from the capital market. Green bonds are an attractive proposition for investors considering environmental, social and governance (ESG) factors or sustainable and responsible investing (SRI) in their investment strategy. The long-term competitiveness of green assets, liquidity associated with the bond market and low operational risk have made green bonds an appealing debt investment avenue for institutional investors, both domestic and foreign.

- **YieldCo** - As the renewable energy investments are long-term and low-risk investments with predictable cash flows, corporates in this sector use yield generating financial instruments to raise debt capital. YiedCo is a long-term financial instrument that enables infrastructure developers to monetize their assets by pooling multiple projects under a single entity (trust). A trust issues units against the underlying completed or revenue-generating assets. YieldCo allows for direct investment by investors in completed and revenue-generating assets, under-construction projects and securities of infrastructure companies. As YieldCo is listed on the stock exchange, the regulation allows a reasonable exit option to its investors. It is also a viable alternative for investors for funds that are looking for liquidity. It is well-positioned as a dividend-paying hybrid instrument for investors with long-term investment horizons with a need for stable cash flows and moderate capital appreciation. Mostly, corporates use this to lower the overall cost of capital by monetizing assets in the operational stage with low risk. It also reduces the unsystematic risk through diversification by pooling multiple small projects.

5.2.2 Equity Financing

Institutional equity investors allocate assets to different sectors based on their strategic allocation policies (Hertrich, 2013). The growth prospects of the renewable energy sector can add significant performance premium and diversification benefits to their portfolios. It can act as a natural hedge against fossil-generated power utilities. Due to increased environmental scrutiny and long-term climate risks (Mercer, 2015) [5], power producers face stranded asset risk and deteriorating financial performance, thus, eroding company profitability. The resulting shift in expected returns from carbon-intensive industries to green sectors warrants rebalancing the institutional investors' portfolios and calls for a change in their asset allocation in favor of green infrastructure, including the renewable energy sector (CPI, 2018) [6]. Equity investment is also strategically suitable for long-term investors, and they are more exposed to long-term risk factors such as climate change, environmental regulation and innovation. Active exposure to the renewable sector or the sectors potentially profiting from climate change risk can improve portfolio performance and address long-term risks (Mercer, 2019) [7]. On the back of strong renewable energy commitments by governments and corporates, the renewable energy sector is expected to witness healthy growth in the coming decades. Equity investors can, thus, allocate a portion of renewable energy to their equity portfolio during the expansionary stage that will allow them to capture the additional upside from equity investment (Mckinsey 2014; Mercer. 2015) [8].

- **Listed Equity -** Listing the company in the stock market is an efficient way of raising capital from the financial market as corporates can access capital from a wide variety of investors, including insurance and pension funds, foreign investors, investment funds and retail investors. Even if pension and insurance funds usually buy and hold equity securities for a long time, they prefer to purchase liquid securities. Foreign investors and investment funds that cannot conduct due diligence and spend resources in structuring required to invest in the private market can also easily invest in the listed company. Retail investors can easily participate in the stock market.

- **Unlisted Equity** – Unlisted equity investments are made either through direct investments or indirectly through unlisted funds. Direct investments involve high transaction cost entailing investment appraisal and due-diligence and deal sourcing cost. Indirect investments occur through unlisted funds, partnerships with local players and co-investment opportunities with other investors. The leading investors in the unlisted companies are strategic investors, private equity players, development finance institutions, sovereign wealth funds, foreign pension funds and asset management companies. The strategic investors are corporates, particularly having a presence in the energy business, who want to enter the renewable energy business by investing in other private companies. These corporates with sufficient cash balance do not wish to fund their carbon-emitting energy businesses and invest in new renewable energy companies. In addition, private equity players also invest heavily in renewable energy companies as the sector is growing at a rapid rate. They look at the exit option when listing securities or selling their stakes to other private equity players. Development financial institutions such as IFC and ADB also invest in unlisted companies to promote carbon-mitigating sectors. Large institutional investors such as pension funds, asset management companies, and sovereign wealth funds are also increasingly interested in buying stakes in renewable energy companies. They want to diversify their energy portfolio and leverage the sector's growth opportunity.

5.3 Why Project Finance

Developing and financing a utility-scale renewable energy project is a complex process and involves the participation of several stakeholders. For the project to be economically viable, the financing structure must allow for optimum risk sharing, allocation and mitigation. It must offset the incremental transaction costs and time incurred in structuring a legally independent project vehicle by entailing significant countervailing benefits. As compared to a corporate finance structured project investment, a project finance structured approach is preferred due to the following reasons:

a) Risk sharing and allocation

The development, construction and operation of a renewable energy project expose the parent sponsor to several risks. These risks may include political risk (expropriation of project assets) or force majeure risk (natural disaster, war or civil disturbance). It may also include market risk (change in the long-term market outlook of the project) or demand risk (inability of the off-taker to perform contractual obligation). As the capital expenditure involved in these projects is very high, realizing any of the risks can lead to volatility in cash flows and contaminate the sponsor's balance sheet (Minton and Schrand, 1999) [9]. Due to high incremental distress costs and absence of risk management apparatus, the company may forego the opportunity to invest in risky projects. A project finance structured investment involves a change in organizational form rather than the use of financial instruments for risk management (Esty, 2004) [10]. In other words, the risks and returns are distributed more efficiently through contractual agreements among those who have the specialized ability to assume the specific project risks.

b) Underinvestment avoidance

Underinvestment is typical of a company that becomes risk-averse, fearing the negative impact of borderline capital expenditures on its cash flows. It usually occurs when capital providers have asymmetric information about assets in place and investment opportunities (Myers and Majluf, 1984) [11]. Inability to raise debt on favorable terms reduces the corporate debt capacity of highly leveraged firms and makes them vulnerable to underinvestment. In project finance structures, the need to value assets in place is eliminated by virtue of the design of the financing structure (Shah and Thakor, 1987) [12]. Due to the separation of current assets and potential investment opportunities, information asymmetry is reduced, and the firm's ability to borrow cheaply is enhanced.

c) Lower cases of agency conflicts

There is a strong correlation between the conflict of interest among debt/equity holders and distortions in investment decisions (Parrino and Weisbach, 1999) [13]. The shareholders (principal) delegate decision-making to a company's management (agent) in an agency relationship. A conflict is inevitable when the objectives of managers who control investment decisions and cash flows do not converge with the interests of the shareholders. If ownership and control are not separated, managers are likely to favor lower risk over higher returns. Project finance incentivizes risk-taking and investment in projects with a positive net present value, leading to the transfer of wealth from debt holders to equity holders.

d) Reduced cost of financing projects

In a traditional corporate financing structure, lenders do not provide funds to a project *per se* but to the borrowing company based on its proven track record and strength of its balance sheet. The debt capacity is limited, and often it becomes difficult for the lender to ensure the use of funds for its intended purpose by the borrower. Due to the assured off-take and guaranteed repayments, a project finance structure allows a higher gearing ratio. The contractual arrangements among project partners allow the availability of non-recourse debt on favorable terms. High leverage also lowers the sovereign risk by making it unviable for the government to confiscate the project. This is unlike conventional financing that increases a project's sovereign risk as a high hurdle rate makes the government feel that the project sponsors are earning profits at the cost of society (Wells and Gleason, 1995) [14].

Utility-scale renewable energy projects are characterized by large development costs. Due to their capital-intensive nature and long exposure period to risk, they are highly sensitive to the financing structure. Complex legal agreements define how the financing will work, with different entities assuming specific risks. A robust appraisal and risk evaluation are necessary for ensuring the viability and bankability of the project. When cash flows don't build up as planned, this documentation acts as the second line of defense. It is, therefore, imperative for the lenders to identify the parties involved in the developmental, financing, implementation and operation stage and bind them in a tight contractual framework.

5.4 Structuring the Project

Project finance differs from corporate finance in terms of structure, contractual bundle and complex documentation. The success of any project depends on the validity and enforceability of the contractual structure in its true spirit. This section provides an understanding of the roles of key project parties and the documentation and contractual structures that bind them.

5.4.1 Project Parties

The key project parties involved are:

- ***Project Sponsors***: Also referred to as promoters or developers, project sponsors are responsible for moving the project from the conceptual to the developmental stage. They play a crucial role in setting up a special-purpose vehicle, mobilizing finance, identifying and recruiting the managerial talent to implement and run the project, and are the principal equity holders of the project.

- ***Special Purpose Vehicle (SPV)***: The SPV is set up to implement and operate the project. It is the counterparty to all the contracts negotiated and executed with project parties and supervises the project construction through an engineering, procurement and construction (EPC) contractor and operations through an operations and maintenance (O&M) contractor. It is incorporated in a jurisdiction that allows for the legal ownership and operation of the project.

- ***Project Lenders***: The lenders provide debt to finance the project's construction and could be banks and other financial institutions or investors in securities like bondholders. Usually, the debt is disbursed by a consortium of banks led by a lead bank that monitors the commissioning of the project and its operations until the repayment is complete. They do not interfere in the day-to-day operations of the SPV. The lenders sign a substitution agreement with the sponsors and SPV, which allows them to step in and affect the conversion of debt into equity, pledge of sponsor's equity, or change of management structure. They can resell the equity to a third party that can carry forward the project profitable.

5.4.2 Contractual Parties

The key contractual parties involved are:

- ***EPC Contractor***: The engineering, procurement and construction contractor is responsible for the building and commissioning the project as per the agreed specifications within a definite cost and time frame. It procures the engineering skills and equipment and subcontracts parts of the assignment to other contractors to construct the project facility. The EPC contract defines the various milestones to be achieved, performance standards, payment schedule and penalties.

- ***O&M Contractor***: The contractor is responsible for operations and maintenance of the project facility and achieves the performance parameters defined in the O&M contract. It provides managerial skills and experience in line with the industry best practices and is usually eligible for a performance bonus (or penalty) if the actual performance exceeds (fall short of) the criteria defined in the contract.

- ***Suppliers***: The suppliers play a critical role in the project development stage and provide the specialized equipment and technology required to set up the project facility. They exercise significant bargaining power with the EPC contractor. The suppliers of raw materials do not exercise much power in renewable energy projects, as the commissioning and operation of the facility depend on the ability to harness the forces of nature.

- ***Off-takers***: The off-taker is the party that is obligated to purchase the project output through a long-term agreement. For renewable energy projects, these contracts are usually in the form of power purchase agreements. The building up of future cash flows and commercial success of the project depends on the soundness of the off-take agreement and counterparty's creditworthiness. This is because the project SPV is directly exposed to the credit risk of the off-taker, and if the latter defaults, the SPV will have to enter into a new off-take agreement at prices lower than the one specified in the initial agreement.

5.4.3 Documentation

The relationship of each of these contractual parties with the project SPV is defined by the underlying documentation. It plays an important role in establishing the contractual bundle for SPV. For a lender, documentation serves as the primary evidence in case of a dispute with the borrower. It makes the terms and conditions for operational performance legally binding on the contractual parties and provides a way to initiate penal action against the defaulter.

In a project finance structured project, there are two types of documents – project documents and financing documents. Together, they are called transaction documents of the project. The project documents include the concession agreement, shareholder's agreement,

EPC contract, O&M contract, power purchase agreement, fuel supply and fuel transportation agreement, etc. The financing documents include the loan agreement and inter-creditor agreement. Each of these agreements is discussed as follows:

- ***Concession agreement***: This is the first agreement signed between the project SPV and government/public body granting the former the right to develop the project. It makes the project site available to the developer for a fixed tenure, usually the project's useful life, and specifies the termination rights after the concession period or force majeure closure. The concession agreement mitigates political risk by laying down the terms and conditions for any direct agreement of the government with the SPV in the form of a state support agreement (SSA).

- ***Shareholder's agreement***: It is the agreement among all the shareholders of SPV. It establishes the shareholding pattern of the project vehicle and representation of the shareholders in the management. This agreement protects bankers against default by a shareholder by defining the latter's exit process. It provides the right of first refusal to other shareholders when an existing sponsor wants to sell its equity stake to an outside bidder. In simple terms, it ensures that the other shareholders get the first right to buy the equity stake.

- ***EPC contract***: This contract is signed between the SPV and EPC contractor. It defines the roles and responsibilities of the EPC contractor in designing, procuring, constructing, testing and commissioning the project facility in a definite period and at the cost specified in the contract. It also mandates the minimum performance parameters to be achieved by the contractor failing which he is liable to pay liquidated damages, also called defects liability.

- ***O&M contract***: This contract is signed between the SPV and O&M contractor. It defines the roles and responsibilities of the O&M contractor in ensuring the successful operation of the project facility and its maintenance as per industry best practices. It provides a safeguard against operating and performance risks. Accordingly, the contractor is awarded/penalized for exceeding/underachieving the performance standards defined in the contract.

- ***Power purchase agreement***: The project cash flows are dependent on the purchase and evacuation of the power generated by an off-taker. In the case of renewable energy projects, the power purchase agreement is usually in the form of a 'take-or-pay' contract. It places a contractual obligation on the off-taker to make periodic payments for the power output at a pre-determined tariff and specified payments even if it does not require the output for a particular time. The agreement can be terminated only by mutual consent.

- ***Loan agreement***: This agreement governs the financing relationship between the SPV and project lenders. It defines the purpose of the loan, covenants to ensure the intended use of funds and repayment schedule. The repayment schedule is usually based on the timeframe for the commercial operation of the project. As the projects involve long gestation periods, the interest rates are generally floating rates pegged to an external benchmark. The loan agreement also lays down the arbitration procedure to be followed by the contracting parties in the event of default.

- ***Inter-creditor agreement***: Utility-scale renewable projects are capital intensive and are usually financed by a consortium of lenders through debt syndication. In this process, several participating lenders come together in the form of a contractual agreement and mutually decide the debt covenants. To prevent unilateral action by any single lender and, at the same time, preserve its rights against the borrower, the agreement defines the rights and responsibilities of each lender.

The following exhibit illustrates the contractual linkages among the various project parties in a typical project finance structure. At the center of this structure are the project SPV and concerned government authority. A concession agreement binds them. The shareholder's agreement governs the constitution of the project vehicle. Lenders come together under an inter-creditor agreement to provide funds to the SPV, which is contractually obligated to safeguard the interests of the lenders under a loan agreement. The SPV enters into an EPC contract with an EPC contractor to construct and commission the project facility. Similarly, it executes an O&M contract with an O&M contractor for the operation and maintenance of the facility. Finally, the off-taker and the project vehicle are bound by a power purchase agreement that ensures the generation of cash flows from the project.

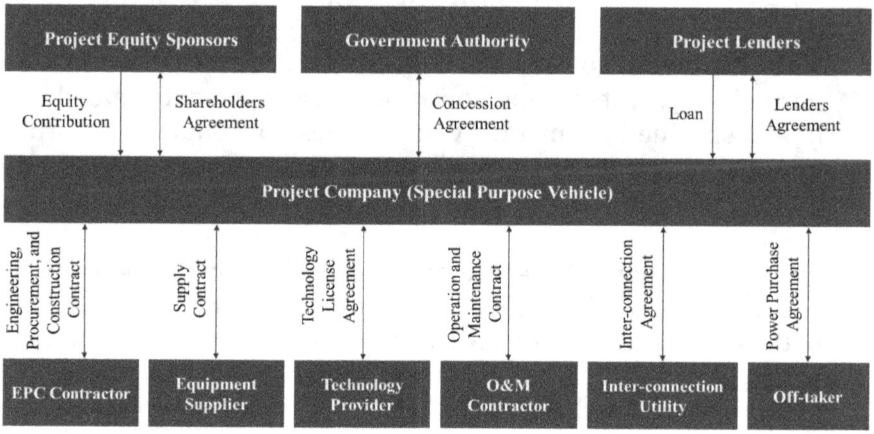

Exhibit 5.1 Typical finance structure for a renewable energy project.

5.5 Financing at Different Stages of the Project

Renewable energy projects can be broken down into three stages: development, construction and operational. There are unique risks attached to each of these stages. As most of the risks are front-loaded, the riskiness of the project decreases with each progressive stage.

a) Development Stage - The risk is the highest at the development stage as the project has to overcome several barriers. The key activities involved in this stage are land purchase, site approvals, obtaining social and environmental clearance, securing contracts from a distribution company and designing the project. This is a cumbersome and time-consuming process and increases the lag between this stage and operation stage of the project. The development stage does not require significant capital expenditure but requires commitment fees to bid for projects if it is a public sector project. Sponsors of the projects typically do not need external capital at this stage. The development stage is a high-risk and high return stage, and it involves a long gestation period, which is not suitable for low-risk investors. The usual investors at this stage are project developers, investment funds, or venture capital funds, and, in some cases, equipment vendors who have the required risk appetite. The primary motivation of these investors is to increase the value of the project for possible re-sale and design an exit strategy to realize their gains from the project investment. They, thus, create a portfolio of projects across a region to diversify the individual project risks.

b) Construction (Greenfield) Stage - This stage requires a heavy amount of capital expenditure - eighty to ninety percent of the project's lifetime expenses. The key activities at this stage are site preparation, construction, final equipment procurement, equipment installation and connection of modules to the grid. The construction duration of a utility-scale renewable project is relatively short, in the range of three to eighteen months compared to thermal power plants. The key risks associated with this stage are related to the timely completion of construction and commissioning. The cost of the project could escalate due to construction delays or cost overruns. The construction stage is a moderate risk and returns stage. To generate high returns, developers and private equity investors have a strong willingness to invest at this stage of the project.

c) Operation (Brownfield) Stage - The operation stage starts after the end of the construction stage. An independent engineer certifies that the project facility has been built to the specifications of the EPC contract and completed all required performance tests (Raikar, 2020) [15]. The operations phase of the project usually lasts for twenty-five years or until the useful life of assets. As compared to the development and construction stages, the risks at the operational stage are low. The primary risk involved at this stage is resource risk, off-taker risk and maintenance cost. This stage is suitable for investors seeking low-risk and low-return. This category of investors comprises credit funds, insurance companies and pension funds interested in investments with a tenor of fifteen years or even up to the useful life of the project assets. These investors seek stable returns from already de-risked or built power plants or low-risk projects and accept the illiquid character of transactions of renewable energy assets.

5.6 Trapping Cash Flows

For a typical renewable energy project, the cash flow predictability stems from two aspects:

a) Revenue: On the revenue side, predictability stems from three key factors - generation, demand and price. Zero cost and low volatility of inputs, proven technology and performance guarantees from equipment suppliers together make renewable energy production fairly predictable. The energy generation risk is further reduced due to the availability of insurance companies to manage resource risk. The risk associated with demand is eliminated by long-term legal power purchase agreements that ensure fixed tariffs or increasing tariffs over the contract period. The term of a power purchase agreement usually varies between 15-25 years in line with the expected useful life of a renewable energy project, thus, making revenue generation reasonably predictable.

b) Cost: in most cases, renewable energy projects do not have recurring capital expenditures (except projects with integrated battery systems, in which the battery needs to be replaced every 5-8 years depending on its usage). Therefore, the operational and maintenance expenditure involved in the project is low and is reasonably predictable. For instance, for a ground-mounted solar project, depending on the size and tariff structure of the project, the operational and maintenance costs range between 5%-20% of the revenue. Any volatility/unpredictability in costs can be further reduced by entering into a long-term O&M contract with a service provider capable of managing such a risk.

The terms of the loan are influenced by the repayment profile of the project, such that the cash outflows are aligned with the inflows. The repayment profile could be back-ended, front-ended, ballooning or equated. It is, therefore, critical for SPV to determine the extent of predictability of cash flows generated by the project and their sufficiency with regard to the repayment obligations – both debt and equity providers.

All transactions of SPV are routed through a trust and retention account (TRA) that serves as the top of a metaphorical waterfall to trap the project cashflows (Groobey et al., 2010) [16]. As the cash flows down the waterfall, it is siphoned off into segregated secured accounts at each different level as per the pattern agreed in a disbursement agreement. These withdrawals usually occur on a monthly or quarterly basis. Any funds remaining at the bottom of the waterfall are paid to the equity sponsors of the project. A typical TRA with a waterfall mechanism is structured in the following way:

The first level of disbursement is used to pay the costs incurred in design, construction and commissioning of the project facility, operation and maintenance expenses, any other approved administration cost, etc. The second level of disbursement is to the lenders to pay loan fees and expenses, interest payments and principal payments (in this order). The third level of disbursement is used to fill a debt service reserve account segregated for the purposes of servicing debt in times of lower project revenues. In this account, two to three-quarters of debt installments are kept as a cushion against default. The fourth level of disbursement is referred to as a "cash sweep," in which the lenders are repaid outstanding principal with a certain percentage of the excess cash remaining after the operation of the three waterfall levels above. The fifth level of the waterfall is used to fill one or more reserve accounts, often designated for future major maintenance or other purposes. Usually, once the reserve account is filled with the required funds, no amounts are out at this level. The sixth level of the waterfall is used to repay the subordinated debtholders or bondholders, if applicable. Finally, the seventh level of the waterfall allows for cash remaining after amounts have been removed at the higher levels to be paid to the equity sponsors of the project in the form of dividends. The waterfall mechanism is shown in the exhibit below:

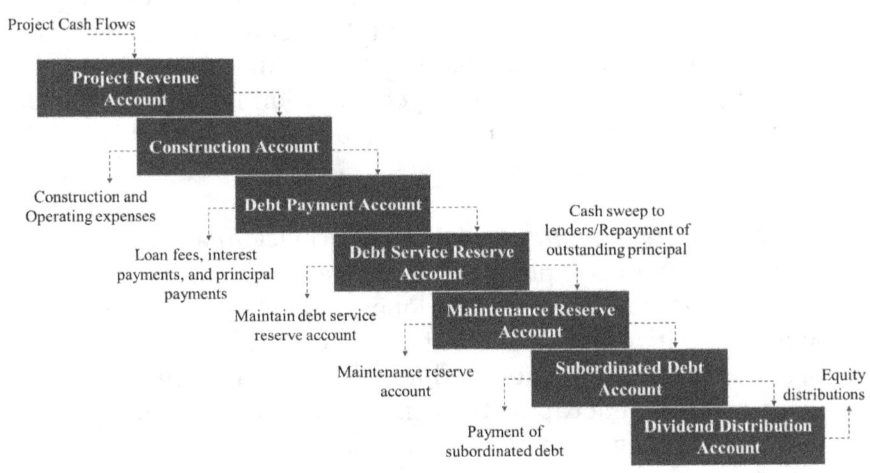

Exhibit 5.2 Waterfall mechanism.

5.7 Risk Management

The identification, assessment, allocation and mitigation of risks associated with the project is at the core of risk structuring. It is necessary to gauge the impact of these risks on project cash flows and create a tight contractual structure to mitigate them. In addition to contractual agreements, lenders place stringent debt covenants or double up risk coverage through a guarantee, indemnity or insurance. The risk management process usually starts with the identification and classification of risk from the perspective of the project parties, external environment or based on the phase of the project.

5.7.1 Risk Identification

These risks are discussed as follows:

- ***Market risk***: For a renewable energy project, the sustainability of tariffs and long-term market outlook can impact its viability. It usually emanates from the lowering of entry barriers as a consequence of policy change or the emergence of technologically superior substitutes. It can lead to variability in the cost structure of the project.

- ***Demand risk***: This risk arises largely due to the limited number of off-takers participating in the project. In most cases, the off-taker is a public entity with the wherewithal to evacuate the power generated and transmit and distribute it to the end consumer. It is highly vulnerable to political interference, and the project gets involved in prolonged litigation in the event of default by the off-taker. To meet the debt service, lenders bind the off-taker in a take-or-pay contract.

- ***Asset liability mismatch risk***: In a project finance structured project, the debt component is as high as 75-80%. Banks usually have short-term liabilities, but longer-term loans. This asset-liability mismatch creates a risk for the lenders. So, banks usually prefer to lend for short or medium-term. Usually, the project developer raises debt capital through the issuance of bonds, primarily targeting long-term institutional investors.

- ***Force majeure risk***: This is an uncontrollable risk, usually referred to as the 'Act of God.' It includes natural disasters like earthquakes, war, civil disturbance, an outbreak of a pandemic, etc. In projects involving only private sector participants, the government's expropriation is also termed force majeure risk. This risk is usually mitigated by using insurance products.

- ***Political risk***: The risk emanating from a change in policy action of the government or local authorities is termed political risk. It includes the risk of expropriation of government facilities by the government, prolonged delays in obtaining necessary approvals for commissioning the project, political activism, regulatory and tax risk, currency inconvertibility risk, etc. This risk can be mitigated by defining arbitration procedure in the concession agreement, using political risk insurance products or by including government as an equity sponsor in the project.

- ***Forex risk***: This risk is typical of projects that have a high component of foreign debt in their capital structure. It arises largely due to exchange inconvertibility - inability to convert local currency into foreign currency when servicing debt - and exchange rate risk due to a depreciation of local currency due to the loan becoming expensive. Forex risk can be mitigated by using derivative products like currency swaps or having a balanced mix of local and foreign currency-denominated loans.

In addition to these risks, the project is vulnerable to specific risks that can be discovered and underwritten in the loan agreement. These risks include cost overrun due to implementation delays, costs escalation due to inflation, management risk arising due to sudden departure of key personnel, use of untested technology and delays in procurement of critical project equipment.

5.7.2 Risk Assessment

The next logical step in risk management is, therefore, risk assessment. Risk assessment is a process of qualitative assessment of risk based on the probability of occurrence of the possible risk factor and the potential economic damage it can cause to the project in the event of its occurrence. All the risks can be arranged in a 2x2 matrix and mitigated by allocating them to a suitable counterparty. In this matrix, the risk is termed high if both the factors are high and low if both the factors are low. Anything in between is termed as medium severity risk. Risks with high severity can negatively impact the project cash flows if not mitigated while structuring the project. This is shown in the exhibit below:

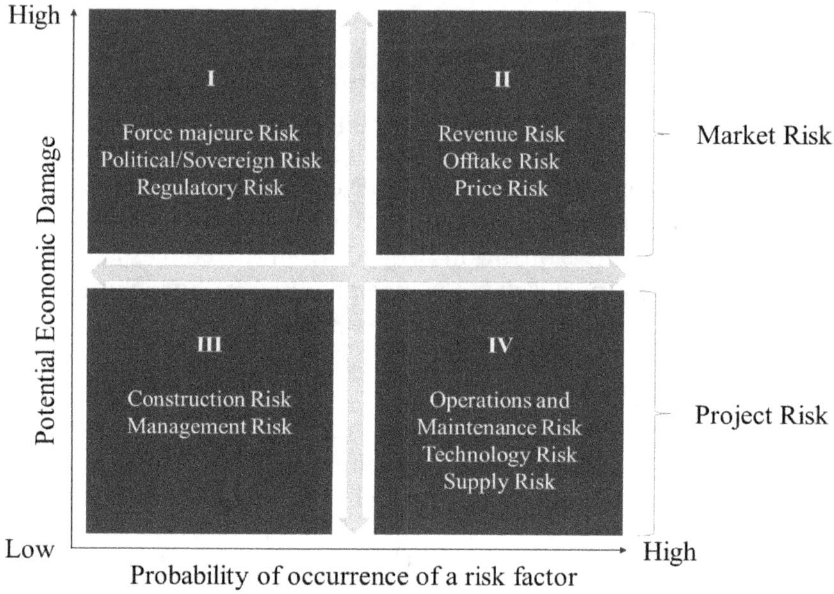

Exhibit 5.3 Risk classification matrix.

5.7.3 Risk Mitigation

Once identified, each of these risks can be mitigated by designing risk management and mitigation strategy. Each risk should be allocated to the party that exercises control over that risk, and accordingly, a counterparty contract should be executed. The contract should be legally enforceable. The following exhibit summarizes the different risks and mitigation strategies associated with renewable energy projects:

Risk	Description	Mitigation
Purchase of land at the project site	Risks associated with the suitability of the project site are environmental pollution, security vulnerabilities, geological factors, etc.	This risk can be mitigated through thorough due diligence by the SPV; making the state/national government an equity sponsor in the project to remove regulatory bottlenecks; transferring risk to EPC contractors through a counterparty agreement.
Social and environmental	The social and environmental risks are related to environmental clearance, just compensation, and rehabilitation support to displaced communities.	This risk can be mitigated by ensuring the inclusion of necessary clauses related to rehabilitation and compensation in the EPC contract.
Project design	The project design is related to output/performance specifications.	This risk can be mitigated by including milestone-based payment criteria in the contract and specifying stringent penalties for non-compliance.
Construction	Risks are related to the adverse effect on the construction of project facilities due to labor disputes, strikes, lack of quality standards, IPR breaches, cost overruns, etc.	This risk can be mitigated by including performance reward criteria in the EPC contract for compliance and penalties for overruns.

Completion	Risks are associated with time and cost overruns and the inability to commission the project on the scheduled commercial operation date.	This risk can be mitigated by making the contractor liable for delayed obligation damages in the contract.
Performance	Risks are associated with meeting output specification metrics and costs.	This risk can be mitigated by making provisions for improvements/ modifications in the facility post-completion.
Resource/Input	Risk is associated with the non-availability of necessary input raw materials for the project operation.	This risk can be mitigated by allowing it's pass through to the contractors for higher fees.
Demand	Risk is associated with the limited number of off-takers participating in the project.	This risk can be mitigated by binding the off-taker in a take-or-pay agreement.
Maintenance	Risk is associated with the maintenance of assets to appropriate standards.	This risk can be mitigated by entering into an O&M agreement with an appropriate service provider.
Force majeure	Risk is associated with unexpected events (Act of God) that are beyond the control of contracting parties.	This risk can be mitigated by purchasing insurance products designed specifically to mitigate the loss of revenue and damages due to force majeure events.
Currency and interest rate	Risk is associated with fluctuation in exchange rates, depreciation of the local currency, etc.	This risk can be mitigated by using suitable hedging instruments.

Insurance	Risk is associated with the non-availability of particular types of insurance products to underwrite non-discovered risks.	This risk can be mitigated through the thorough due diligence of the project in consultation with an insurance advisor.
Political	Risk is associated with excessive government interference or expropriation of project assets.	This risk can be mitigated by purchasing political insurance products or by making the government an equity partner in the project.
Regulatory	Risk is associated with frequent changes in laws or the prevailing tax rates.	This risk can be mitigated by using suitable insurance products.
Inflation	Risk is associated with a persistent increase in project costs above expected under the power purchase agreement.	This risk can be mitigated by using suitable hedging instruments.
Strategic	Risk is associated with an exodus of project sponsors, conflict of interest, changes in shareholding patterns, managerial incompetence, etc.	This risk can be mitigated by transferring strategic decision-making to the project SPV and specifying mechanisms to handle such conflicts in the shareholder agreement.
Disruptive technology	Risk is associated with unanticipated displacement of current technology that changes the cost dynamics of the project.	This risk can be mitigated by adding relevant clauses to the EPC contract to take full advantage of new technologies during the project's construction phase.

Exhibit 5.4 Different types of project risks and mitigation strategies.

5.8 Conclusion

The global energy order is undergoing a profound transformation, from fossil fuels to one based on renewable energy to mitigate climate change risks. According to the Intergovernmental Panel on Climate Change (IPCC), a low-carbon transition in line with limiting warming to 1.5 degrees Celsius would require a massive upscaling of supply-side investments, about USD 3.5 trillion annually between 2016-2050. At present, the key challenge to the realization of the globally agreed climate objectives is the ability of the private sector to mobilize sustainable and adequate financial resources for various sectors, including renewable energy. Only public finance is not enough to meet the amount of capital required for renewable energy to meet the desired goal, hence, there is an urgent to accelerate private capital to the sector. In this context, financers need to have a clear understanding of the economics of renewable energy projects to make informed decisions.

The character of the renewable energy projects is different from conventional energy – these require large upfront investments than the latter, but their operating and maintenance costs are significantly less. At the same time, there are several unique risks attached to renewable energy projects, and if not properly managed, the projects can become commercially unviable. The private financiers need to identify the risks profile of renewable energy projects at different stages to make an informed investment decision. At the same time, the renewable energy developer needs to understand the possible sources of financing at various stages of the project. The developer must identify the possible risks of the project and deploy an appropriate risk mitigation strategy. As a primer, this chapter covers financial models, mode of financing and investment strategies for the renewable energy business.

References

1. Frankfurt School-UNEP Centre/BNEF, (2017). "*Global trends in renewable Energy Investment.*"
2. Nevitt, Peter K., and Frank J. Fabozzi. *Project Financing*. Euromoney Books, 2021.
3. Jena, Labanya Prakash, and Chavi Meattle. (2020) "Directing Institutional Capital to India's Renewable Energy Sector." In Innovation,

Technology, and Market Ecosystems, pp. 109-128. Palgrave Macmillan, Cham.
4. Hertrich. C, (2013), "*Asset Allocation Considerations for Pension Insurance Funds: Theoretical Analysis and Empirical Evidence.*" Springer.
5. Mercer (2015), "*Investing in a time of climate Change.*"
6. Jena, L.P.; Meattle, C.; Shrimali, G. (2018) Getting to India's Renewable Energy Targets: A Business Case for Institutional Investment; CPI Report: San Francisco CA, USA.
7. Mercer (2019), "*Investing in a time of climate Change – The Sequel.*"
8. McKinsey & Co., (2014), *From indexes to insights: The rise of thematic investing.*
9. Minton, Bernadette A., and Catherine Schrand. *"The impact of cash flow volatility on discretionary investment and the costs of debt and equity financing."* Journal of Financial Economics 54, No. 3 (1999): 423-460.
10. Esty, Benjamin C. *"Why study large projects? An introduction to research on project finance."* European Financial Management 10, No. 2: 213-224.
11. Myers, Stewart C., and Nicholas S. Majluf. (1984) *"Corporate financing and investment decisions when firms have information that investors do not have."* Journal of Financial Economics 13, No. 2: 187-221.
12. Shah, Salman, and Anjan V. Thakor. (1987) *"Optimal capital structure and project financing."* Journal of Economic Theory 42, No. 2: 209-243.
13. Parrino, Robert, and Michael S. Weisbach. (1999) *"Measuring investment distortions arising from stockholder–bondholder conflicts."* Journal of Financial Economics 53, No. 1: 3-42.
14. Wells, Louis T., and Eric S. Gleason. (1995) *"Is foreign infrastructure investment still risky?".* Harvard Business Review 73, No. 5: 44-53.
15. Raikar, Santosh. "SA (Ed.) (2020) 13-Renewable Energy Finance in the International Context." Renewable Energy Finance, Academic Press: Cambridge, MA, USA.
16. Groobey, Chris, John Pierce, Michael Faber, and Greg Broome. (2010) *"Project finance primer for renewable energy and clean tech projects."* Wilson Sonsini Goodrich & Rosati: 29.

6

THD Measurement and Simulation of Grid-Connected Solar PV and Wind Integrated Renewable Energy System

Venkata Suresh Babu Chinni

Faculty of Applied Sciences, AIMST University, Semeling, Malaysia

Abstract

This chapter takes us through different objectives to make the grid-connected hybrid system work effectively. In this chapter, double-fed induction generator-controlled wind energy system integrated with a solar PV plant is connected to the grid. The observed non-linear variation of solar irradiation and temperature was controlled by the MPPT controller, which tries to maintain the maximum power condition at almost all instants. In this study, the incremental conductance type MPPT method is employed for solar PV plant, whereas, in the case of a wind farm, the use of back-to-back connected VSC incorporated in the DFIG maintains unity power factor for different wind speed variations. As energy availability from renewable sources is intermittent, the harmonics at the load side are inevitable. These can be reduced by using a 3-Level VSC with suitable PWM techniques to some extent. Herein, three different cases to find the best way to integrate the wind and solar PV energy systems into the Grid have been considered. MATLAB/SIMULINK was used to perform simulations with THD measurements for three different cases.

6.1 Introduction

Energy demand is rising rapidly across the world due to modernization. At present, 80% of the world power is facilitated from the traditional energy sources unsafe to the atmosphere [1-6]. In recent surveys, it is reported that by 2050 worldwide, energy demand will have tripled. Hence, there is every need to utilize the available renewable energy sources. Solar PV and wind energy incorporated systems are better choices for fulfilling future energy needs. Despite the benefits of sustainable energy integrated systems and their reaction to underbalanced states, their performance gives preliminary results in defective grids [7-9]. Various control techniques have been recommended to advance the strength of the same for superior performance in anticipation of an inoperative grid [4-5].

The primary goal of this work is to demonstrate modelling, analysis, and total harmonic distortion (THD) estimation of grid-connected combined solar PV and wind energy systems. Three situations were considered in this research for the study of power system stability and power quality.

Case 1: Wind energy system linked to the grid
Case 2: Solar PV array system linked to the grid
Case 3: Integration of solar PV and wind energy system as resources to the grid.

6.2 Grid Connected Solar PV-Wind Energy System and its Features

Owing to their abundant nature and continually decreasing investment prices, solar PV and wind energy sources are the enticing electrical energy options. Wind and solar PV energy sources have a global installed capacity of more than 539 GW and 405 GW, respectively, by 2017 [2]. The climate conditions affect the wind and PV energy sources, primarily caused by wind speed and solar irradiation. In the range of cut-in and rated wind speed values, the output power is noted to be non-linear and non-zero [4]. The generated power is not consistent in certain situations, and it may be less than the power demand [3-6]. As a result, it must be combined with other energy sources to ensure a continuous power delivery to end-users. As illustrated later, the output power of the solar PV cells is computed based on the cell's I-V characteristics. Because V_{oc} and I_{sc} are primarily dependent on the solar irradiance and PV cell temperature, the solar PV power plants provide fluctuating power. As a result, standalone solar PV power plants are not recommended for sensitive loads that require a constant and uninterrupted power supply.

6.2.1 Wind Energy System Linked to Grid

In Figure 6.1, a nine-megawatt wind farm with six wind turbines each rated at 1.5 MW and linked to a 25 kV distribution grid supplies power to a 120 kV grid over a 25 kV feeder spanning over 5 km and 14 km.

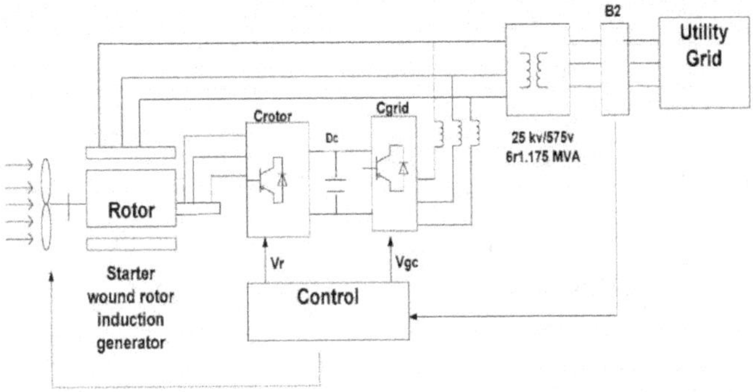

Figure 6.1 Grid Connected wind energy system.

The stator windings are directly connected to the transmission grid, while the rotor is at variable frequencies through the AC/DC/AC converter. The wind turbines utilize the double fed induction generators (DFIG) to create the rotor voltage and torque. DFIG extracts the maximum energy from the wind, even for low wind speeds, by optimizing the turbine speed, while minimizing mechanical stress on the turbine. The doubly fed induction generator-based WES mainly consists of a generator, RSC, wind turbine with the drive train system, DC-link capacitor, GSC, coupling transformer, pitch controller and protection system. The stator of the doubly-fed induction generator is directly connected to the grid, and rotor is coupled to the mains through an AC/DC/AC converter that only handles about 25-30% of the total power to carry out full control of the generator. The principle of variable speed generator was obtained with the combination of DFIG and four-quadrant AC/DC/AC VSC provided with IGBTs. The AC/DC/AC converter system contains an RSC and a GSC linked back-to-back by a DC link capacitor. RSC is employed to change the machine excitation and electro-magnetic torque. The operation of the power converter in bidirectional power modes enables DFIG to work either in sub-synchronous or in super-synchronous operating modes. Table 6.1 provides the parameters of DFIG [5].

Table 6.1 DFIG parameters

Rated power	6*1.5 = 9 MW
Stator frequency	60 Hz
Stator nominal voltage	575 V
Stator resistance	0.023 pu
Rotor resistance	0.016 pu
Stator inductance	0.18 pu
Rotor inductance	0.16 pu
Inertia constant	0.685 pu
Nominal DC bus voltage	1150 V

A fourth-order state-space model indicates the variation of the doubly-fed induction generator through a synchronously rotating reference frame (d-q frame).

Control of DFIG-based WECS

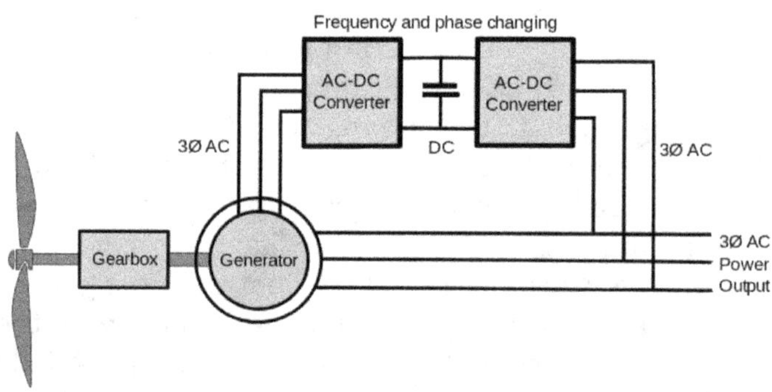

Figure 6.2 Block diagram of DFIG-based WECS.

A control block diagram of DFIG-based WECS with two levels of control is illustrated in Figure 6.2. One of the levels is the maximum level which is WECS optimization. In this, the speed of the wind turbine is adjusted such that it can obtain the optimum wind power. This control level is mechanical system control. Second is the lower-level electrical system control, i.e., torque and reactive power control. When compared to the electrical control system, the mechanical control method is slower. Figure 6.3 shows the Cp curves for wind turbines. The variation of turbine power, tip speed ratio (lambda, λ) and Cp values with wind speed is made evident in Figure 6.4. It can be seen that, for a wind speed of 15 m/s, the output power of the turbine is 1 per unit, the pitch angle is 5.7 degrees, and the speed of the generator is 1.2 perunit.

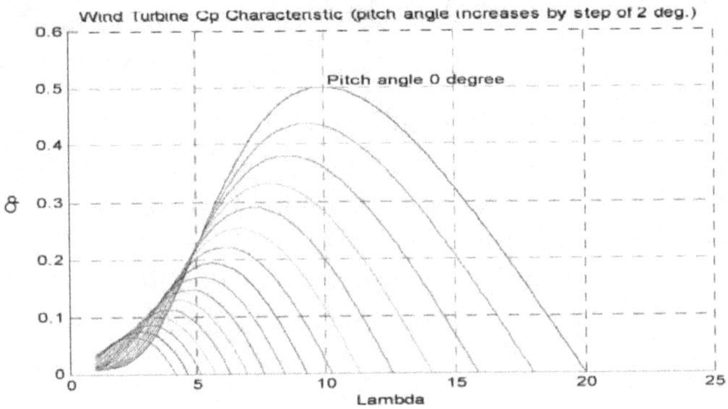

Figure 6.3 Pitch angle for wind turbines (increases for each step of 2 degrees).

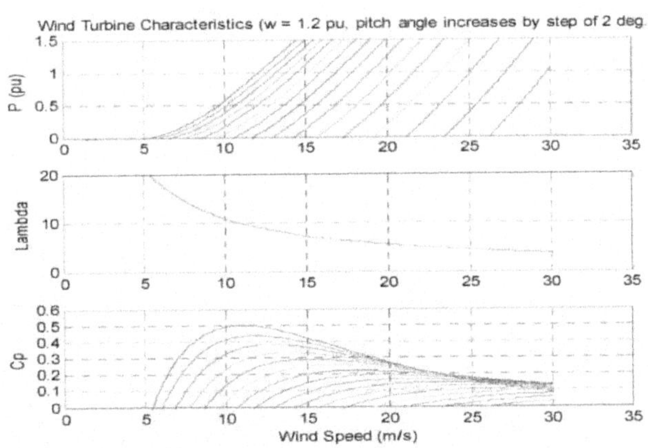

Figure 6.4 Wind turbine C_p characteristic (ω=1.2 pu; the pitch angle increases by a step of 2 degree).

6.2.2 PV Array System Connected to Grid

Figure 6.5 depicts a 100-kW array connected to a 25-kV distribution grid. At 1000 W/m^2 sun irradiance, the PV array can produce up to

100 kW power. The voltage is raised from PV natural voltage (272 V DC at maximum power) to 500 V DC by a boost converter working at 5 kHz, with the MPPT controller enhancing the switching duty cycle of the converter. The incremental conductance plum is used in this controller. The incremental conductance with integral regulator technique is used in this controller. At unity power factor, the VSC transforms the available 500 V DC to 260 V AC.

A capacitor-bank advantage of 10 kVAr to filter harmonics is created by voltage source converter (VSC), a 100 kVA 260 V/25 kV three-phase coupling device. The utility grid model (25 kV classification feeder + 120 kV equivalent transmission arrangement), B1 (grid side bus) and B25 (load side transport) are provided in Figure 6.6. A PV array consists of abundant series and parallel linked PV modules, while a distinct PV module contains various series-connected solar cells.

The general equation defining the I-V characteristics of an ideal cell is:

$$I_d = I_{sat}\left[e^{V_d/V_T} - 1\right] \quad (1)$$

$$V_T = k\frac{T}{qQ_dN_{cells}} \quad (2)$$

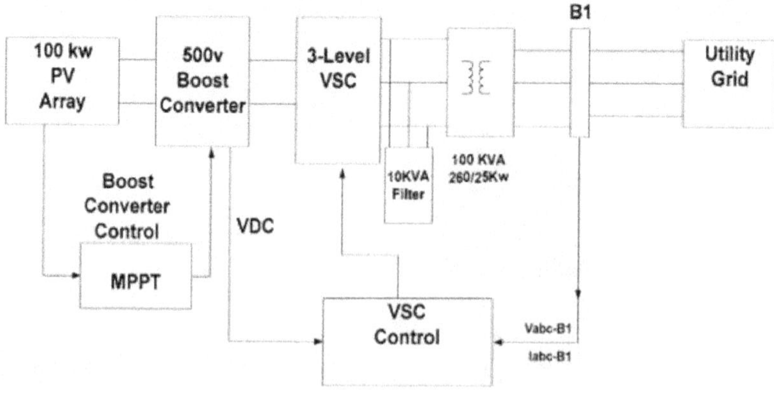

Figure 6.5 PV connected to grid.

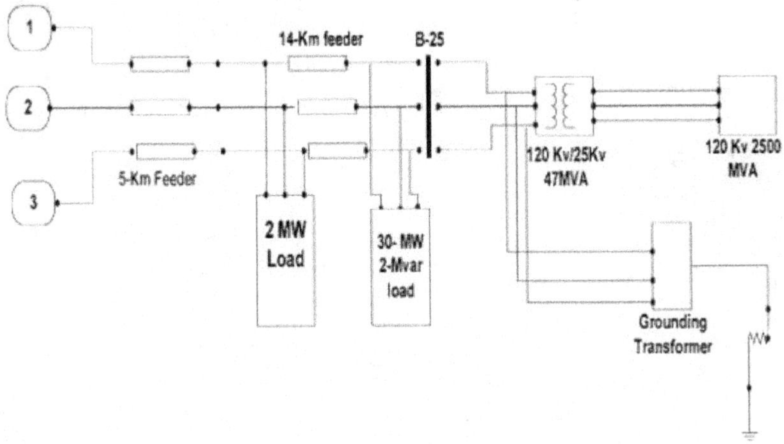

Figure 6.6 Utility grid.

where,

I_d represents the diode current.
I_{sat} represents saturation current of the diode.
Vd represents voltage of the diode.
VT represents the temperature voltage.
k represents the Boltzmann's constant value.
q represents charge of the electron.
Q_d represents quality factor the diode.
N_{cells} represents the number of series connected cells per module.
I_{ph} represents the single module light photo-current generated.

The value of current generated rises in direct proportion to the solar irradiation on the module's surface as the number of parallel module strings rises. For a PV array with sixty-six parallel strings (Npar) consisting of five modules (Nser) each, Equation (1) is altered to:

$$I_{darray} = I_{satarray}[e^{Vd/Vtarray} - 1] \quad (3)$$

where, I_{darray} is the aggregated diode current for the PV array, I_{sat} array is the aggregated diode saturation current, and V_{Tarray} is the aggregated temperature voltage. These values, along with the increased photo-current values, are calculated as:

$$I_{pharray} = I_{ph} \text{ X } N_{par} \tag{4}$$

$$I_{satarray} = I_{sat} \text{ X } N_{par} \tag{5}$$

$$V_{T}\ array = V_{T} * N_{ser} \tag{6}$$

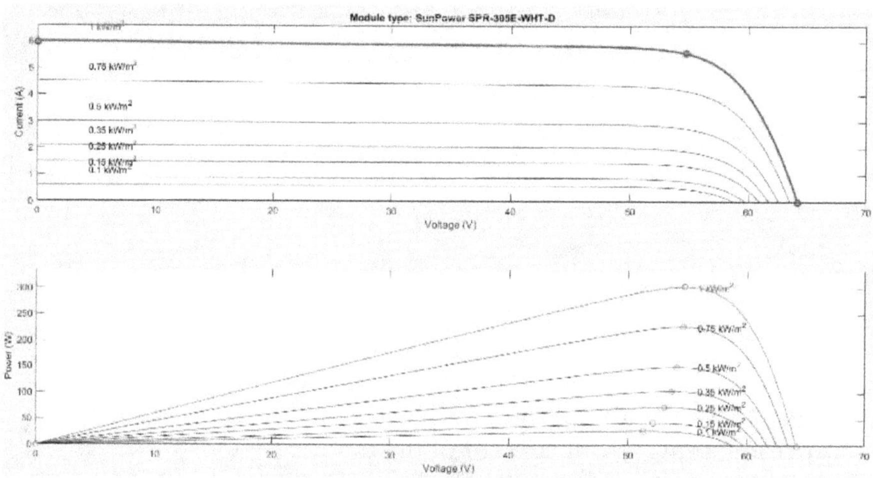

Figure 6.7 I-V and P-V characteristics of PV array.

In Figure 6.7, the dotted plots indicate deferential values of input irradiances (25 °C, 100 W/m²). The circle underlined on the plots represents the extreme or maximum point, and the dotted plots indicate reverent of input irradiance values (25 °C, 100 W/m²). Table 6.2 shows the parameters of the sun powered SPR350-WHT PV array.

The determined power for a single PV array (watt) is therefore given by:

$$P_{array} = N_{par} \ X \ N_{ser} \ X \ P_{mp} \tag{7}$$

From (7)

$P_{array} = 100.7\,\text{kW}$

Table 6.2 Sun powered SPR-305-WHT PV array specifications

Parameter	Variable (units)	Value
Number of cells in series	N_{cells}	96
Number of series linked modules per string	N_{ser}	5
Number of parallel strings	N_{par}	66
Maximum power	P_{mp}(W)	305.2
Maximum power voltage	V_{mp}(V)	54.70
Maximum power current	I_{mp}(A)	5.58
Voltage value of open circuit	V_{oc}(V)	64.20
Voltage Value of Short-circuit current	I_{sc}(A)	5.96
Voltage value of series connected resistance	R_s(Ω)	0..0380
Voltage value of parallel connected resistance	R_p(Ω)	993.5
Diode saturation current	I_{sat}(A)	3.1949e-8
Light generated photo-current	I_{ph}(A)	5.9602
Diode quality factor	Q_d	1.3

6.2.3 Combination of Solar PV System and Wind as Hybrid Sources to the Grid

Figure 6.8 shows the integration of a 100 kW PV array (BI Bus) and a 9 Mega Watts wind farm with six 1.5 Mega Watts wind turbines coupled to a 25 kV grid (B2 Bus).

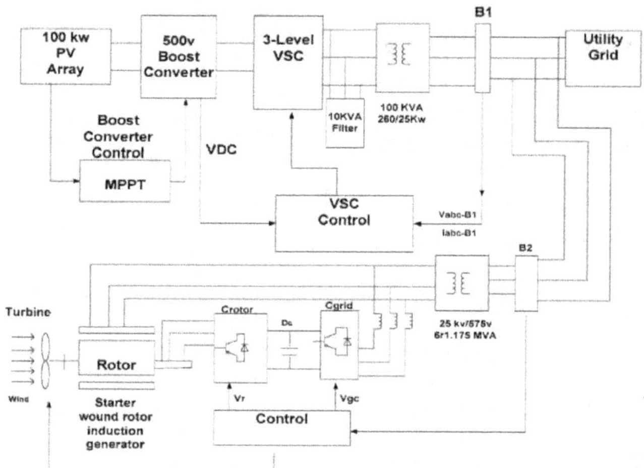

Figure 6.8 Integration model of PV system and wind as hybrid sources to the grid (Mohod, 2018).

6.3 Simulation

Wind Energy System Linked to Grid

The model wind speed is kept constant at 15 m/s during the first run. The DFIG (dual-fed induction generator) wind farm produces 9 MW power. The turbine speed is 1.2 times the synchronous generator speed. These control the wind turbine's reactive power at 10 Mvar and regulate the DC voltage at 1150 V.

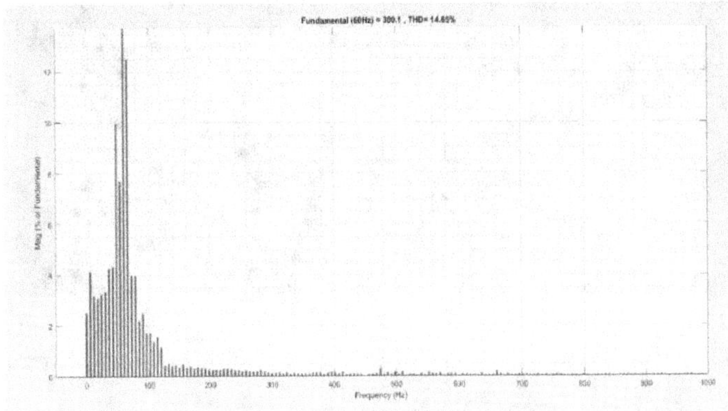

Figure 6.9 Grid side bus (B_2) THD.

PV Array System Connected to Grid

The pulses to the boost and VSC converters are restricted from t = 0 s to t = 0.05 s when the model run for 0.7 s. The PV voltage is similar to 321 V open-circuit voltage. The three-level bridge acts as a diode rectifier and charges the DC link capacitors to over 500 V. The boost and VSC converters are de-blocked at t = 0.05 s. Vdc = 500 V is used to manage the DC link voltage. The boost converter's duty cycle was fixed at D = 0.5, and the sun's irradiance was set to 1000 W/m^2. At t = 0.25 s, the steady-state PV voltage of 250 V is reached as a result. The PV array output power is 96 kW, with a maximum capacity of 100.7 kW at 1000 W/m^2 irradiance. MPPT is allowed at t = 0.04 s. The MPPT regulator begins regulating PV voltage by varying the duty cycle to extract the maximum amount of power. When the duty cycle is at D = 0.0453, the maximum power value (100.7 kW) is obtained. PV mean voltage is 274 V at t = 0.5 s, based on the PV module specifications of 273.5 V. Various irradiance adjustments are applied from t = 0.5 s to 0.7 s to demonstrate the MPPT controller's optimal performance. An index is shown in Figure 6.9. PV array voltage boost is 500 V at t = 0.2, and the index value is 0.85.

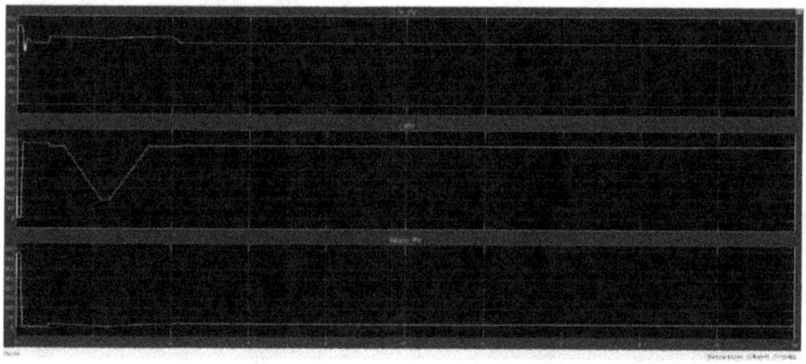

Figure 6.10 Voltage and current from PV array under different irradiations.

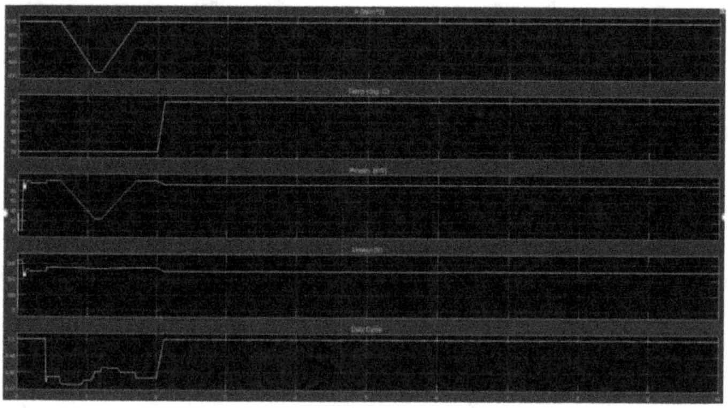

Figure 6.11 Variation of temperature, solar irradiation and duty cycle of boost converter.

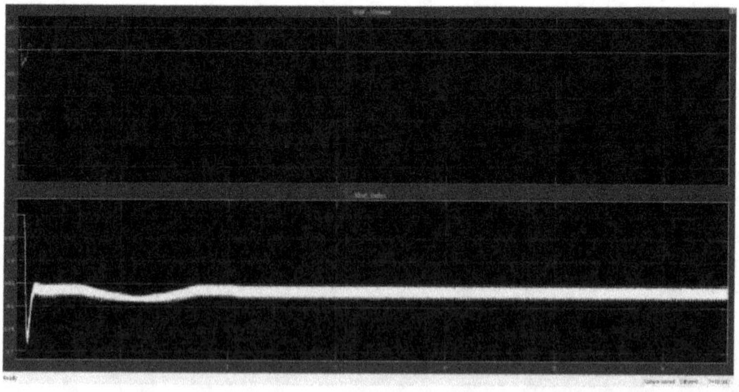

Figure 6.12 V_DC voltage (500 V), Mod-index and id, iq graphs.

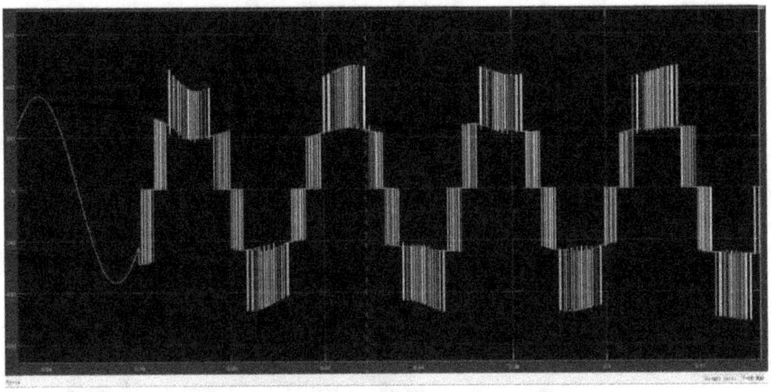

Figure 6.13 Output of three-level VSC.

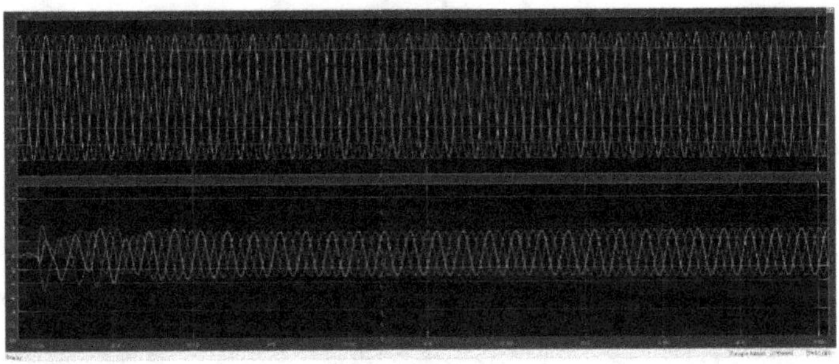

Figure 6.14 Grid side bus B1V_abc, I_abc.

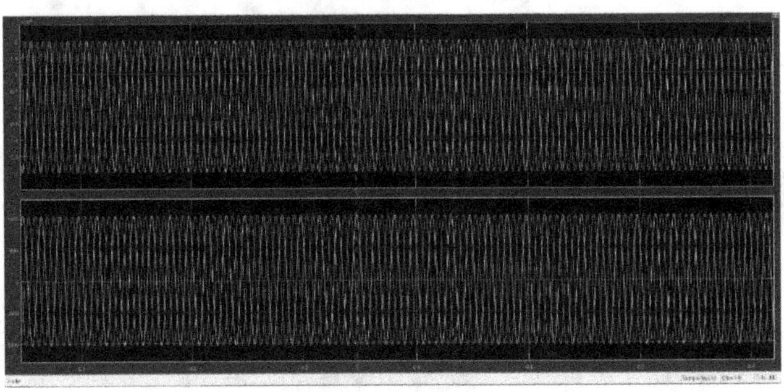

Figure 6.15 Load side bus B_25 Vabc, Iabc.

The voltages and currents of the grid side bus (B 1) and load side bus (B 25) are shown in Figure 6.10-6.15, with the current THD in the grid side bus (B 1 bus THD) as 26.09% and load side bus (B 25 THD) as 0.11%.

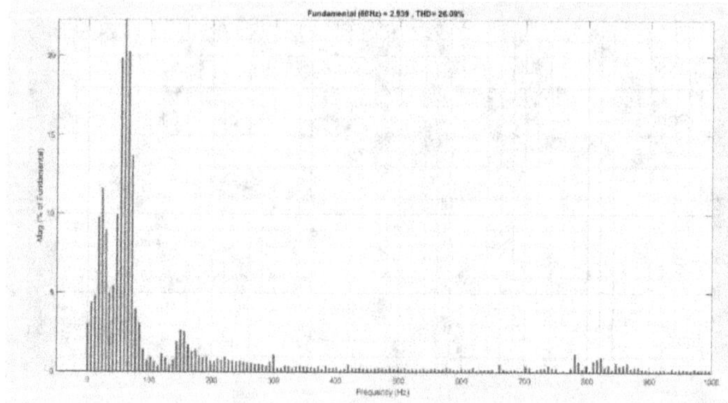

Figure 6.16 THD of Bus 1 (B_1).

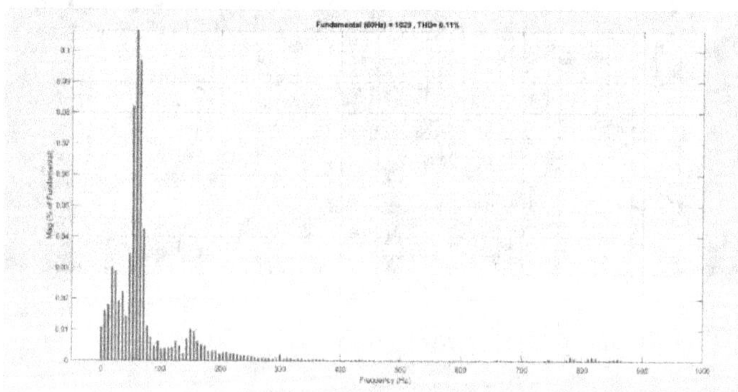

Figure 6.17 THD of Bus 25 (B _25).

Integration of PV System and Wind as Hybrid Resources to the Grid

Simulation study and THD measurement, capacity of current and voltage of B_1bus, B_2 bus and B_25 bus are shown in Figures 6.18-6.20. The THD of current of B_1 bus is 58.64%, B_2 bus is 18.10% and B_25 bus is 0.16%, as shown in Figure 6.21-6.23.

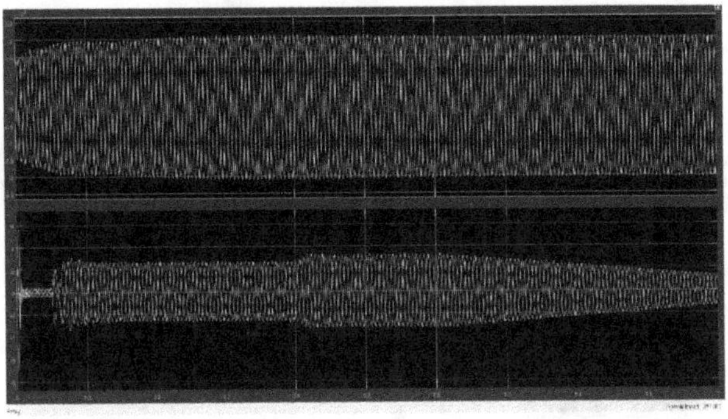

Figure 6.18 Vabc and Iabc of Bus B_1.

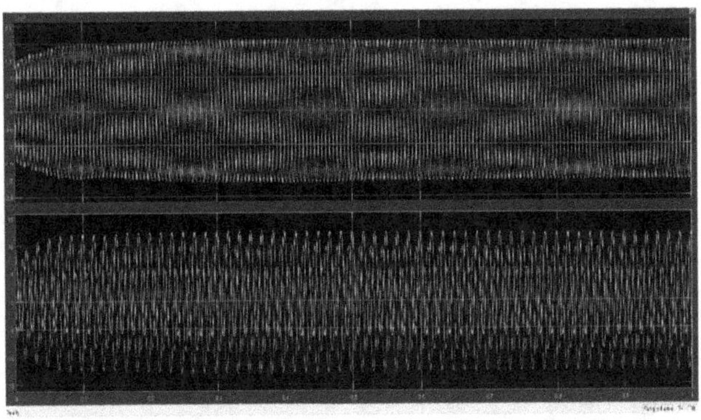

Figure 6.19 Vabc and Iabc of Bus B_2.

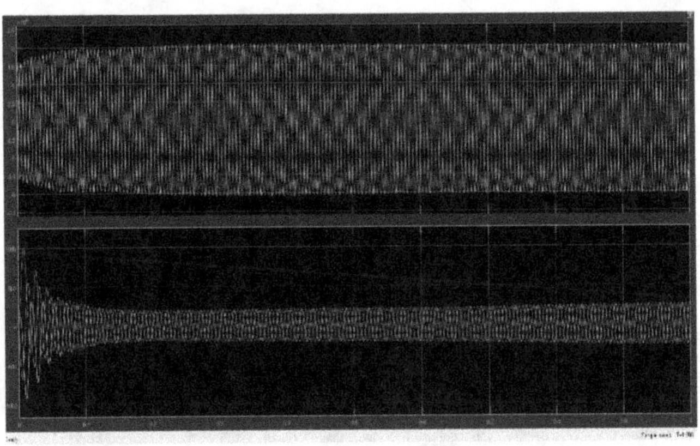

Figure 6.20 Vabc and Iabc of load side Bus B_25.

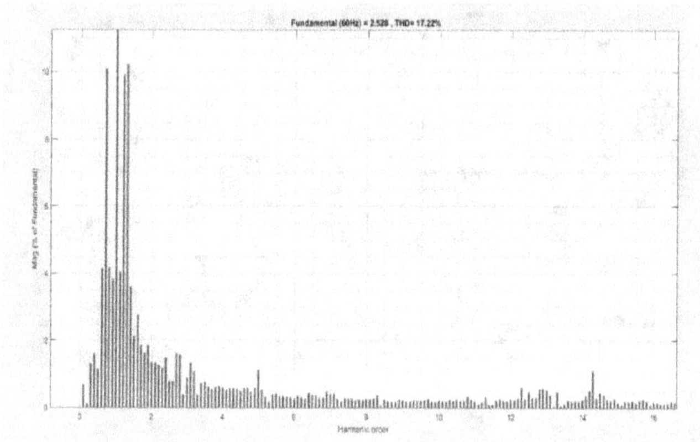

Figure 6.21 THD of BUS B_1.

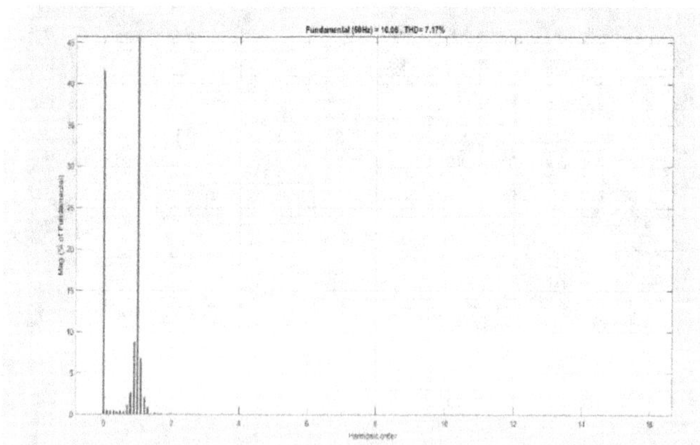

Figure 6.22 THD of Bus B_2.

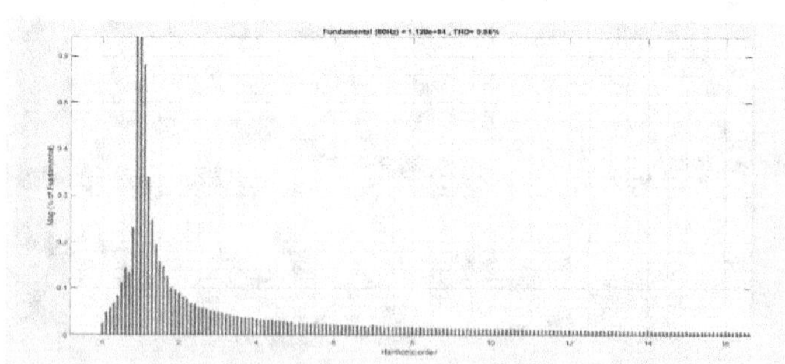

Figure 6.23 THD of load side Bus B_25.

Table 6.3 Comparison of THD values of current for different cases

	Case A	Case B	Case C
B_1 Bus	26.09	-	17.22
B_2 Bus	-	14.65	7.17
B_25 Bus	0.11	5.58	0.86

As seen in Table 6.3, the THD of Case C is the lowest when compared to Case A and Case B. THD decreases from Case A (B 1 Bus) to Case C (B 1 Bus) from 26.09 to 17.22, and from Case B (B 2 Bus) to Case C (B 2 Bus) from 14.65 to 7.17. THD is noted to decrease from 5.58 to 0.86 in B 25 BUS (load side bus) Case B to Case C , however, it is observed to grow from 0.11 to 0.86 in Case A to Case C.

6.4 Conclusion

The DFIG and PV energy transformation system integrated with the grid had been favorably simulated using MATLAB. Three different cases are measured to study the outcomes and determine the present THD. The system has been tested under a variety of PV irradiation conditions as well as boost converter voltages. In comparison to Case A and Case B, Case C exhibits the lowest THD. THD reduced from Case A (B _1 Bus) to Case C (B 1Bus) from 26.09 to 17.22, and from Case B (B 2 Bus) to Case B (B2 Bus) from 14.56 to 7.17. THD reduced 5.58 to 0.86 in the B 25 Bus (load side bus) in Case B to Case C, but improved 0.11 to 0.86 in Case A to Case C.

References

1. Bhardwaj, S. (2018). The future of publicpolicy.
2. Demolli, H. (2019). Wind power forecasting based on daily wind speed data using machine learning algorithms.
3. El-Ghonemy, A. (2012). Photovoltaic Solar Energy: Review. International Journal of Scientific & Engineering Research.
4. Esteban, M. et al. (2010). Methodology to estimate the output of a dual solar-wind renewable energy system in Japan. Energy Policy. 38. 7793-7802.10.1016/j.enpol.2010.08.039.

5. Naresh Kumar et al. (2017). Stability Analysis of Wind Farm Based on DFIG by STATCOM.IJSRD. International Journal for Scientific Research & Development, 5-6.
6. Mohod, S. B. (2018). Hybrid power system with integration of wind, battery and solar PV system. IEEE.
7. Shao, G. N. (2016). Renewable and sustainable energy reviews.139-150.
8. Subarto Kumar Ghosh. (2013). Modeling of PV Array and Analysis of Different Parameters. International Journal of Advancements in Research & Technology.
9. Praveen P., Shaik M. A., Kumar T. S., and Choudhury T. (2021) Smart Farming: Securing Farmers Using Block Chain Technology and IOT. In: Choudhury T. et al. (eds) Blockchain Applications in IoT Ecosystem. EAI/Springer Innovations in Communication and Computing. Springer, Cham.

7

Artificial Neural Network-based Advanced MPPT Algorithm for New Generation Solar PV Systems Under Varying Irradiance

Anubrata Sengupta

School of Illumination Science Engineering and Design, Jadavpur University, Kolkata, India

Abstract

Solar PV has been one of the primary renewable energy sources in the current context due to its typical characteristics and sophisticated operation. Most of the times, the output of solar PV has been observed to be irregular and unpredictable. Due to this, the load end gets stressed most of the times. The power generation via photovoltaics (PV), due to their benefits such as ease of availability, low cost, negligible environmental pollution, lower maintenance tariff, etc., has been gaining popularity compared to other available renewable resources. In order to minimize the drastic effects of changing environmental conditions over the output of the PV system, the maximum power point tracking (MPPT) technique has been adopted in most of the areas. It helps to keep track of the panel's maximum power output to increase the overall energy generation. Easy design, low cost, good performance characteristics with minimal output power variability and the ability to monitor changing conditions easily and quickly are significant features of the MPPT controllers. An MPPT system based on an improved neural network has been proposed and developed in the current study. As compared to the existing software computing technologies and conventional monitoring arrangements, the proposed system has a lower transient and steady-state response. An extensive research has been carried out on a standalone solar photovoltaic system for multidimensional performance analysis. The output has been studied, and considerable changes have been highlighted, followed by necessary explanations.

7.1 Introduction

The government of India supports the idea of self-reliance, thus, setting an ambitious target of meeting the country's rapidly rising energy demand through clean green sustainable resources, rather than fossil fuels which have been used intensively since the '90s. The government has targeted to produce 40 percent of total energy generation through renewable energy or DERs by 2030 [1]. The government has been continuously making an effort to produce 175 GW of renewable energy by 2022 with prevailing solar power. Increasing environmental concerns, higher carbon footprints, declining fuel supplies, and increased energy requirements have directed focus towards an aspirational future based primarily on renewable-driven sources and the need for the hour too [2]. In the solar cell photovoltaic power conversion process, two major breakthroughs have been made. To begin with, light assimilation results in the formation of an electron-hole pair. The gadget's structure isolates the electron and the hole. The electrons go to the negative terminal, while the holes go to the positive terminal. The power is provided based on the distribution of holes and electrons. The perfect terminal voltage and current are set using photovoltaic solar picture displays that are combined similarly, arranged, or mixed. The arranged string configuration allows for a higher voltage level, but the individual photovoltaic cell value constrains current evaluations [3]-[5].

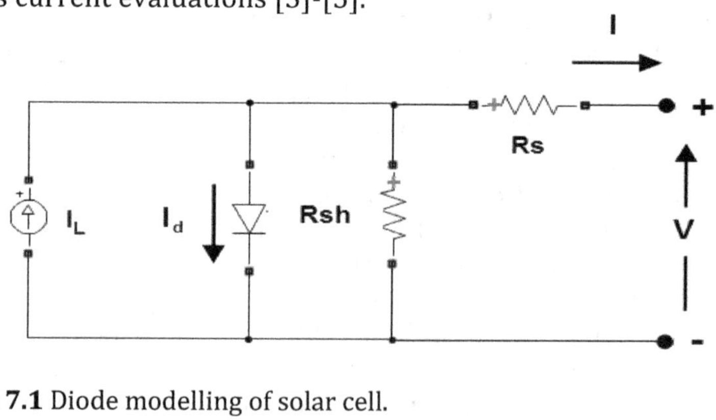

Figure 7.1 Diode modelling of solar cell.

Figure 7.1 depicts the solar cell modelling diode. The solar cell diode model is used to calculate the I-V and P-V characteristics of solar cells. The controlling equations for the diode model are discussed, taking into account the Rs and Rp effects:

$$I = I_{sc} - I_{01}\left[e^{q\left(\frac{V+I.R_S}{kT}\right)} - 1\right] - I_{02}\left[e^{q\left(\frac{V+I.R_S}{kT}\right)} - 1\right] - \left(\frac{V+IR_s}{R_p}\right) \quad (1.1)$$

$$I = I_{sc} - I_0\left[e^{q\left(\frac{V+I.R_S}{nkT}\right)} - 1\right] - \left(\frac{V+IR_s}{R_p}\right) \quad (1.2)$$

where, n is known as the "ideality factor," and the value of the ideality factor is usually determined by solar cell manufacturing technology. A solar cell is a semiconductor that converts sunlight into electricity. Solar cells, for example, will produce electricity using electromagnetic power if the photons' power is high enough to discern electron matches. A single or multi-crystal solar cell produces an electric flux voltage of 0.5 volts [6]. Solar irradiance is the electromagnetic radiation emitted by the sun. The voltage of the solar cell's N/P obstacle layer causes this. The sum of electrons thumped into the conduction band determines the current or amperage of the solar cell. This current is linked to the measurement of solar radiation by the sun. The current of a solar cell can be increased by increasing the solar cell's surface area or by increasing the solar cell's measurement of solar radiation. A solar battery is made up of solar cells. As solar cells are combined, the current stays the same, but the voltage increases at the same time [7].

The solar cells are connected to form a "module" that supplies current and voltage to the system (and therefore power). To frame a 12-volt module, for example, 24 solar cells must be attached in a scheme. A photovoltaic module is also known as a solar cell array [8]. The power is proportional to the current voltage. When the sun shines at a rate of 1000 W/m^2 and the temperature is 25 °C, the power level of a photovoltaic module is commonly referred to as the module's power output. This is an approximation of average sunshine in the middle of a clear summer day. A 15 percent effective 1 m^2 square module will, thus, produce 150 W in the early afternoon. A photovoltaic (PV) exhibit is a set of power-generating photovoltaic modules. A PV display can be made up of only one module, with output ranging from a few watts to several megawatts, depending on the number and output of the modules. The "heap" is powered by the direct current produced by a photovoltaic display [9]. From charging a battery to powering a matching system in a minicomputer to powering a structure or city, batteries are used in a variety of applications. An inverter that adjusts the immediate current in the current should be associated with a PV cluster when it is linked to the utility network [10]. The majority of inverters have a 90% efficiency rating. The advanced inverters generate extremely clean electricity at a constant voltage. Clean power denotes a spinning current that is nearly free of mutilation or harmonics, similar to a sine wave. Via maximum power point control, the aim is to improve the operational efficiency of solar photovoltaic systems. The source impedance and load impedance can be adjusted by changing the duty cycle of the corresponding boost converter, allowing for complete power transfer monitoring from the photovoltaic system [11].

7.2 MPPT

The standard solar photovoltaic models with one and dual diodes are common. The single diode model is less complex, but the dual diode model is more advanced for improving the solar PV modeling performance. On the other hand, double diode displays have a more complicated and mathematical charge in parameter extraction. The solar photovoltaic system shown is in Figure 7.1 in electrical equivalent circuits. The performance equation for a single diode model is as follows.

$$I = I_{PV} - I_0\left(\frac{(V+IR_S)}{nV_T} - 1\right) - \left(\frac{(V+IR_S)}{R_{Sh}}\right) \tag{2.1}$$

where,

I_{PV}	output current of PV module
A_{I0}	diode saturation current, A
I_D	diode current, A
I_{Sh}	shunt Current, A
R_S	series resistance, Ω
Rsh	shunt resistance
V_T	thermal voltage, V
V	output voltage of PV array, V
I	output current of PV array, A
N_S	number of series cell connected
N_P	number of parallel cells connected
K	Boltzmann constant ($1.3806503 \times 10^{-23}$ J/K).

q electron charge ($1.60217646 \times 10^{-19}$ C)
T temperature, ° C.
n fill Factor (ideal=1)

Thermal relation is provided as

$$V_T = KT / q \tag{2.2}$$

The diode current expression is indicated by

$$I_D = I_0(e^{qVd}/n^{kt} - 1) \tag{2.3}$$

The current load expression is given by

$$I = I_{PV} - I_d - I_{sh} \tag{2.4}$$

Shunt current is given by

$$I_{sh} = (V+IRs)/Rsh$$

The current phase equation is calculated by

$$I_{PV} = I - I_0 \, [\exp\,(Vph + RshIsh/\,n) - 1] \tag{2.5}$$

The current equation is given by the reverse saturation

$$I_0 = n_p \, I_{ph} - n_p \, I_{rs} \, [exp\,(KV/n_s) - 1] \tag{2.6}$$

V-I characteristics represent a relationship between the current and voltage in the solar cell under various irradiation and temperature conditions. This curve evaluates the parameters and behavior of certain solar cells.

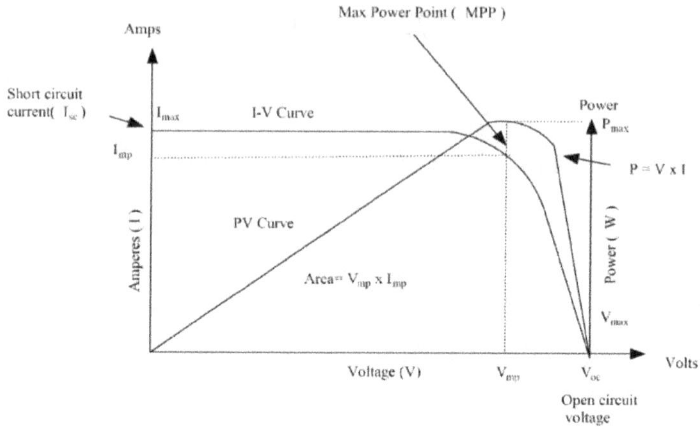

Figure 7.2 V-I characteristics curve of solar cell.

Figure 7.2 depicts a standard solar PV cell that operates at a standard test condition (STC). The characteristic curve demonstrates the relation of voltage and current, which in turn is the result of solar cell power generation. The solar cells are open-circuited and not connected to a load, ensuring that the current is zero, and the cell voltage is at its maximum. When the voltage transverse solar cell is zero, the current through the solar cell is known as a short circuit (ISC).

$$P_{max} = V_{oc} \text{ x } I_{SC} \tag{2.7}$$

$$P_{mpp} = V_{mp} \text{ x } I_{mp} \tag{2.8}$$

where,

$\mathbf{V_{oc}}$	open voltage circuit
$\mathbf{I_{sc}}$	short circuit current
$\mathbf{P_{max}}$	maximum power
$\mathbf{V_{mp}}$	High voltage at the point of service
$\mathbf{P_{mpp}}$	Highest power at the operational condition

The maximum power point monitoring methods are used to obtain the maximum value of power in a solar system, allowing the most reliable and maximum amount of power to be sent from the source to the load. We know that solar radiation and temperature fluctuate during the day, so an algorithm that can monitor the MPP is needed. These can have a major impact on the PV system's performance. It is also claimed that if the system's operating point is not close to MPP, a large number of losses would occur. The "Maximum PowerPoint" is the voltage point at which the power value is at its highest. This point, however, varies with solar irradiance and temperature, and the key challenge is determining the optimum voltage and current points for maximum power under varying atmospheric conditions. The majority of MPPT methodologies depend on PV characteristics such as duty cycle management and the use of a look-up table.

7.3 ANN (Artificial Neural Network)

A human brain is a versatile tool capable of completing even the most difficult tasks. Artificial neural networks are built on a model of the human brain that has been created artificially. It is not feasible to replicate the human brain in its entirety at this time, and there are certain limitations to what current technology can do artificially. It is possible to create condensed artificial neurons and an artificial neural network. It is possible to obtain the interconnections of the neurons. ANNs can be produced in a variety of ways to mimic the brain. The artificial neural network performs admirably when it comes to recognizing basic patterns or solving complex problems. They are artificially used in intelligence analysis due to their exceptional training skills. ANNs are educated using data sets with a large amount of data. For the first time, the neural network is trained with information comprising various types of images. Even if the given data input exceeds the neural network's data collection, the input provided to the network can be categorized and categorized once educated. As a result, the characteristics do not need to be specified directly. The network learns to differentiate between mammalian and non-mammalian species by receiving training from a mammal. The most popular feed, multilayer feeds, is chosen to be implemented using ANN. The outputs are returned based on the input value. An ANN feed is generated in the following way: the layer receives the input, and the output is returned by the output layer and passed through all hidden layers [2,3].

$$Y(x) = g\ (W_n{+}1 \sum_{l=0} W_{1xi}) \tag{3.1}$$

$$g(x) = 1/\ (1 + e^{25x}) \tag{3.2}$$

To obtain an output, it is easy to spread the input via an ANN feeder's network.

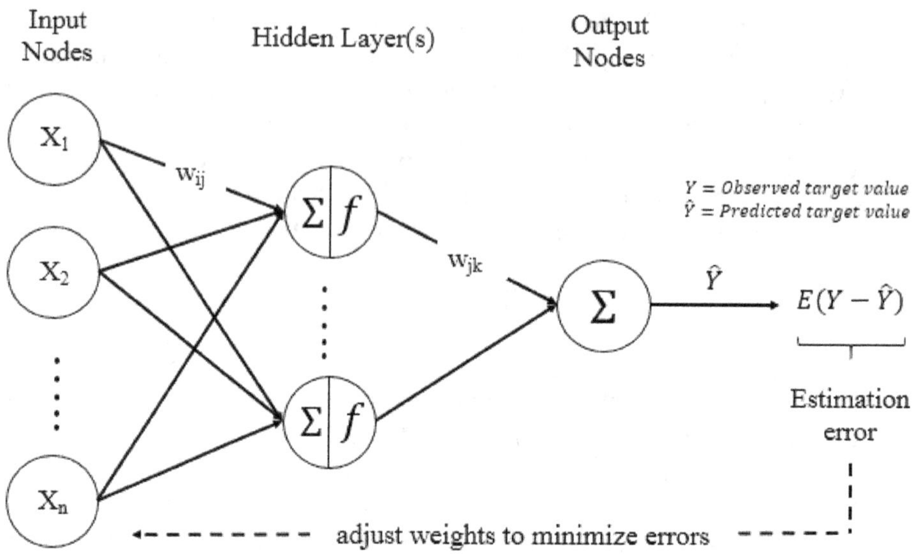

Figure 7.3 Fully linked multi-layer feed to a hidden layer.

The issue occurs when dealing with a network that has links in all directions (as in the brain) and needs to measure output. These loops can be processed with the aid of recurrent networks since links are made within the network. The recurrent networks can code time dependency, but forward networks can better provide feedback on problems that are not time dependent.

Figure 7.3 depicts a multifaceted ANN feed that links all of the neurons in each layer to the next layer. This network is known as a completely linked network, and the ANNs are normally, but not always, fully connected. The mass assigned to different inputs is one thing, and the value inactivation functions are another. There are two parameters to change when training a single ANN. Such an arrangement is impractical and adjusting only one parameter will make the device easier to manage. To solve this problem, a bias neuron is developed. The outcome is still 1, as it is with the bias neuron. The neuronal bias is only linked to the next layer of neurons, not the previous one. The weights of the ANN type are always 1 in an initial state since the product of a partial neuron is always 1. An input layer is the first layer, and an output layer is the last layer. There are three different types of layers. Between the two multi-layer feed forwards, a number of layers are linked. ANN has layers that are called hidden layers. The relation moves only forward from the first layer to the next layer. Feeding forward with multi-layer is a good idea. ANNs are divided into two phases: the first is referred to as the learning process. If a specific input is given during the training process, a specific output is generated. The ANN receiving ongoing training on a collection of training data accomplishes this. The total or combined number of the remaining weights is explicitly applied and processed via the similar activation function value in the second step of implementation. The prejudice neuron's addition aids in the deletion of the activation function's t. If the ANN has been conditioned, the weights just need to be changed. This means that "t" is not removed without the addition of a bias neuron since it is the complete value. The function will show, regardless of the weight value, whether all inputs are zero.

7.4 Proposed Methodology

ANN is based on information collected from the incremental (Inc-Cond) method. The neuron-based artificial network comprises three layers: input, hidden, and output.

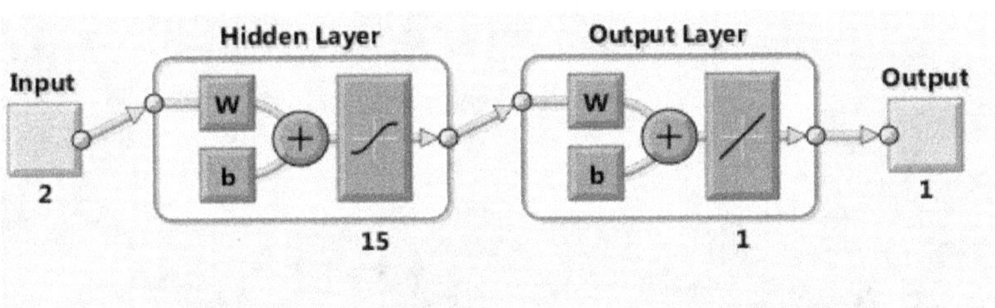

Figure 7.4 Design of neural network.

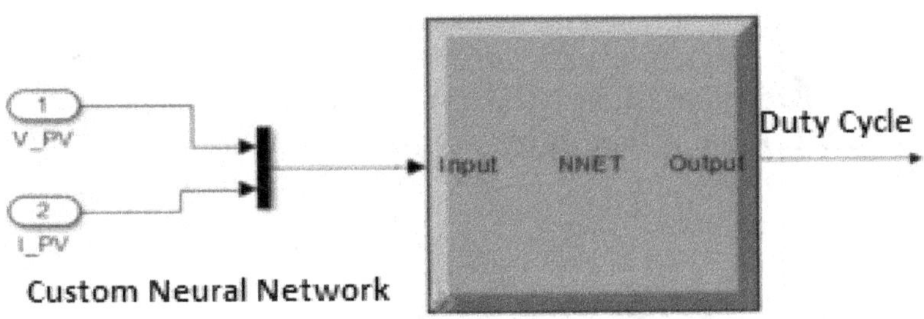

Figure 7.5 Neuron model for MPPT system.

The PV voltage, current and solar irradiance can be used to train ANN or any combination of these temperatures. Neural network learning is carried out by feed-forward weight updating. For this purpose, 200000 data collected from the INC algorithm are used for training.

$$E_{mse} = \sum_{k=0}^{n} = \frac{1}{12}[m(k) - o(k)]^2 \tag{4.1}$$

Propagation of back Levenberg- Marquardt algorithm with PV and PV current is carried out, where m(k) refers to the output measured, o(k) refers to the output required, and N refers to the number of training patterns. The hidden layer contains 15 neurons and is activated tangentially by the sigmoid produces output from the hidden layer, while neurons are trained to output layer linearly. The output layer function is enabled, and the neural network performance function is analyzed with its medium error (MSE).

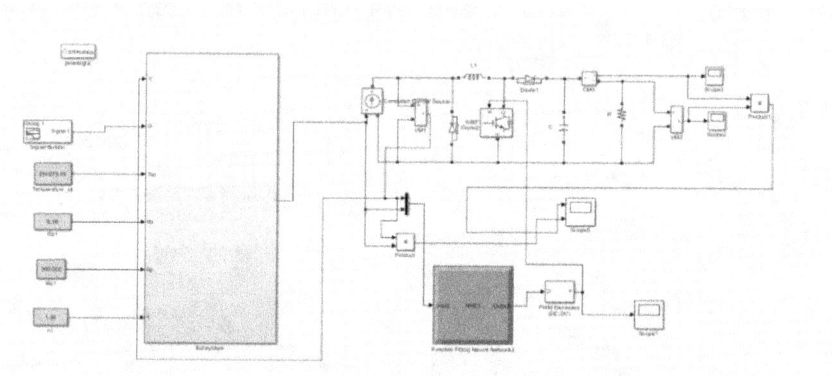

Figure 7.6 Design of MATLAB Simulink model for proposed system.

The benefit of the proposed algorithm is that it takes more time to track MPP. Everyone PV array parameters alter with time, and the neural network must therefore be trained to ensure accurate MPP tracking regularly. Input to the subsystem of the PV panel, i.e., irradiation is varied quantitatively.

7.5 Results and Observations

In this exploration, a soft computing-based charge controller was developed to effectively track the maximum IV power points of PV systems with variable irradiances and complex working conditions, and soft computer-based control systems with enhanced incremental conductance-based trained artificial neural network configuration and duty cycles were controlled.

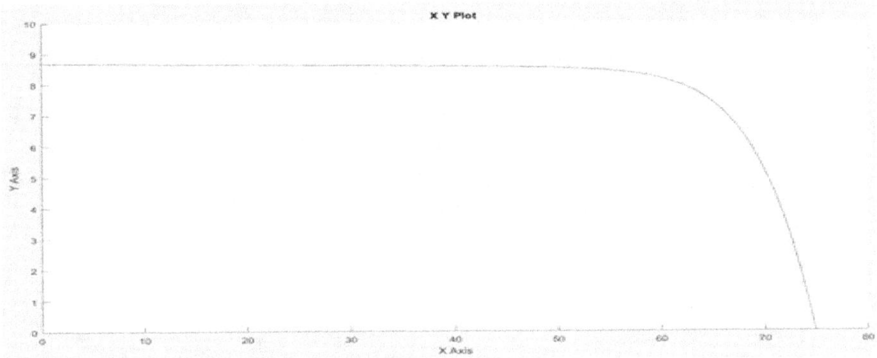

Figure 7.7 V-I characteristics of solar panel.

The PV modules and battery on the storage system and inverter are connected to the load controller in high-energy applications. According to current research, a photovoltaic solar system's operating performance can be improved by using a powerful power converter with a soft-computer MPPT controller.

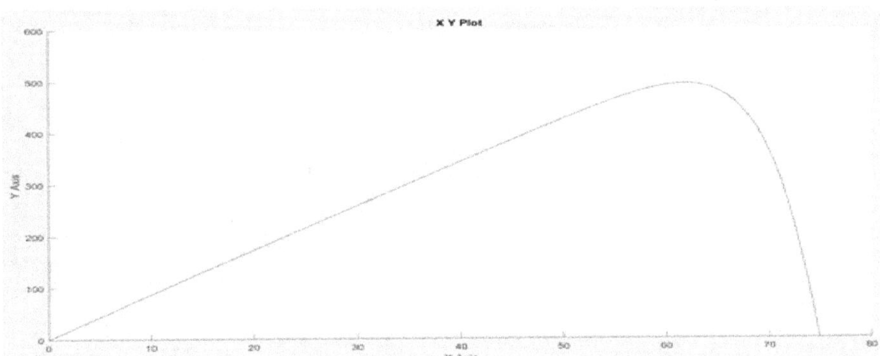

Figure 7.8 P-V characteristics of solar panel.

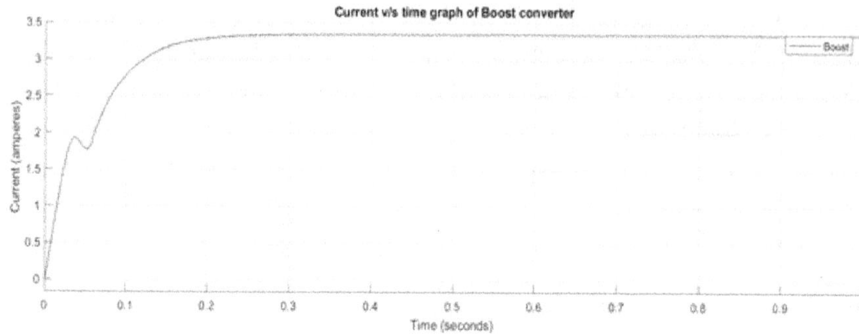

Figure 7.9 Analysis of current for boost converter.

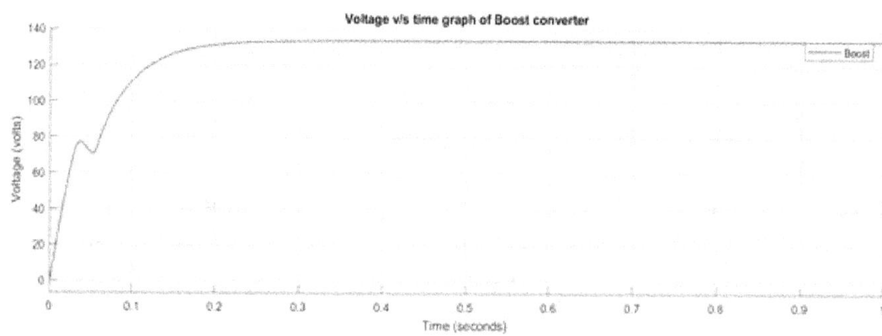

Figure 7.10 Analysis of voltage for boost converter.

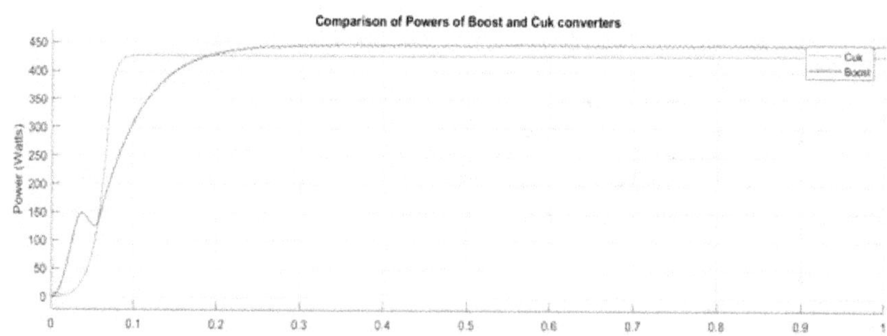

Figure 7.11 Comparative analysis of power for proposed system.

Table 7.1 Analysis of results

MPPT	Power	Efficiency	Ripples
P and O	424 W	84.8 %	0.083
INC	430 W	86 %	0.081
ANN-INC Hybrid (Proposed)	450 W	90%	0.05

The addition of an effective charge controller based on an intelligent maximum energy monitoring system improves the operating efficiency of solar photovoltaic systems. A battery-contaminated charging controller can interact with solar photovoltaic device storage applications and high-power inverters.

7.6 Conclusion

The need of the hour is to harness more energy out of the same installed solar PV unit or system by the implementation of a new effective MPPT algorithm. ANN-INC-based power point controllers were developed during this study to track efficiently with variable irradiance and complex operating conditions. The maximum power point was derived from photovoltaic systems' I-V characteristics, and the soft computer program-based controls with boost configuration and operating cycles were controlled using enhanced ANN methodology. The PV modules and battery on the storage system and inverter were connected to the controller. As can be observed, a photovoltaic solar system's operating performance can be improved by using a powerful power converter with a soft-computer MPPT controller. The addition of an effective charge controller based on an intelligent maximum energy monitoring system improves the operating efficiency of solar photovoltaic systems. The study can be expanded to include the use of a deep learning convolutional neural network for photovoltaic device design and analysis.

References

1. Sriram, M., & Ravindra, K. (2020). Backtracking Search Optimization Algorithm Based MPPT Technique for Solar PV System. In Advances in Decision Sciences, Image Processing, Security and Computer Vision (pp. 498-506). Springer, Cham.
2. Satapathy, Susree Sukanya, and Nishant Kumar. (2019) "Wavelet Mutation based JAYA Optimization Algorithm for Global Maximum Power

Peak Searching for Partially Shaded Solar PV Panel Condition." In 2019 3rd International Conference on Recent Developments in Control, Automation & Power Engineering (RDCAPE), pp. 351-356. IEEE.

3. Raj, Ashish, Manoj Gupta, and Sampurna Panda. (2016) "Design simulation and performance assessment of yield and loss forecasting for 100 KWp grid connected solar PV system." In 2016 2nd International Conference on Next Generation Computing Technologies (NGCT), pp. 528-533. IEEE.
4. Z. Cheng, H. Zhou, and H. Yang, (2010) "Research on MPPT control of PV system based on PSO algorithm," 2010 Chinese Control Decis. Conf. CCDC 2010, pp. 887–892.
5. Ashwini, K., Raj, A., & Gupta, M. (2016) 'Performance assessment and orientation optimization of 100 kWp grid connected solar PV system in Indian scenario'. In 2016 International Conference on Recent Advances and Innovations in Engineering (ICRAIE) (pp. 1-7).
6. Sharma, Rahul S., and P. K. Katti. (2017) "Perturb & observation MPPT algorithm for solar photovoltaic system." In 2017 International Conference on Circuit, Power and Computing Technologies (ICCPCT), pp. 1-6. IEEE.
7. Das T.K., Banik A., Chattopadhyay S., Das A. (2019) Sub-harmonics Based String Fault Assessment in Solar PV Arrays. In: Chattopadhyay S., Roy T., Sengupta S., Berger-Vachon C. (eds) Modelling and Simulation in Science, Technology and Engineering Mathematics. MS-17 2017. Advances in Intelligent Systems and Computing, vol 749. Springer, Cham.
8. T. K. Das, A. Banik, S. Chattopadhyay and A. Das, (2019) "FFT based Classification of Solar Photo Voltaic Microgrid System," 2019 Second International Conference on Advanced Computational and Communication Paradigms (ICACCP), Gangtok, India.
9. A. Banik and A. Sengupta, (2021) "Scope, Challenges, Opportunities and Future Goal Assessment of Floating Solar Park," 2021 Innovations in Energy Management and Renewable Resources (52042), Kolkata, India, pp. 1-5.
10. A. Banik, A. Shrivastava, R. Manohar Potdar et al., (2021) Design, Modelling, and Analysis of Novel Solar PV System using MATLAB, Materials Today: Proceedings, https://doi.org/10.1016/j.matpr.2021.06.226
11. Das T.K., Banik A., Chattopadhyay S., Das A. (2021) Energy-Efficient Cooling Scheme of Power Transformer: An Innovative Approach Using Solar and Waste Heat Energy Technology. In: Ghosh S.K., Ghosh K., Das S., Dan P.K., Kundu A. (eds) Advances in Thermal Engineering, Manufacturing, and Production Management. ICTEMA 2020. Lecture Notes in Mechanical Engineering. Springer, Singapore. https://doi.org/10.1007/978-981-16-2347-9_17

8

Mitigation of Power Quality Issues in Autonomous Solar Photovoltaic Utility Microgrid Employing a STATCOM based Intelligent Controller

Rohan Banik[1] and Pradip Das[2]

[1]*Dept. of Civil Engineering Malda Polytechnic, West Bengal, India*
[2]*Dept. of Electrical Engineering, Ratua Satyendra Nath Bose Government Polytechnic, West Bengal, India*

Abstract

Microgrid has emerged as an efficient autonomous decentralized utility grid to reduce the dependency on the regular electric grid. It also benefits in the reduction of unwanted transient and harmonics and results in overall power quality improvement. This study is essentially focused on the solar PV unit because of its increasing importance and promising outcome despite several inferiorities. It has been observed that a reactive power divergence has created a significant power quality issue in the solar PV-based microgrid systems, which in turn results in significant voltage fluctuation and deterioration in overall power quality. This fluctuation has a considerable impact on both steady-state and transient stability and is sometimes responsible for the inadequate performance of the solar PV unit. Hence, this work is mainly dedicated to compensating the reactive power and to overcome the problems of maintaining a constant voltage profile of a solar photovoltaic-based standalone or hybrid microgrid system. Therefore, this study employs a fuzzy controller and a static synchronous compensator (STATCOM). The obtained results or outcomes reveal that this proposed novel cutting-edge system successfully maintains stability and improves power quality in utility PV-microgrids.

8.1 Introduction

As per the World Energy Outlook report 2020, global electricity demand will grow at 2.1% per year till 2040. The incorporation DERs in the energy sector can only meet this dramatic exponential growth demand. The scarcity of fossil fuels has led to the development of a new energy infrastructure called "smart grid." Thus, smart grid design is based on green energy technologies such as PV, wind, biomass, hydro, etc. Implementing this as an alternative energy generation will benefit in lower transmission losses and improved reliability. Instead of using an individual energy generation, the combination of two or more systems will yield better reliable power. As wind and solar are the more prominent sources, this work formulates a hybrid system using wind and solar. However, the output of both systems depends upon the environmental conditions. So, energy backup should be implemented to increase the security level of the available energy. Hence, battery/fuel cell can be employed as a storage device. Thus, the designed microgrid can be operated in either grid-connected/islanded mode. Apart from the pollution-free environment, the microgrid can create power quality problems due to its random output nature. In a standalone system, when the load is changed, the system's voltage profile gets affected due to power variations. If non-linear loads are introduced, current harmonics are injected into the system. Thus, these power quality issues can be eliminated by introducing a filter combination or by FACTS devices. Conventional LC filter technique utilized for power quality issue results in resonance with the system's impedance [3]. Hence, STATCOM can be incorporated to mitigate power quality issues in MG. It can either absorb/inject reactive power into the network. This regulates power flow and improves power quality aspects of the system. Simultaneously, it will remove the stress imposed by wind energy generators also [4]. In order to regulate the operation of STATCOM, controllers are implemented. As the PI controller exhibits a wide range of stability margins, they are not suitable for non-linear variations of microgrid [11] Hence, various controllers are used for the smooth operation of STATCOM. Icosφ controller was introduced to reduce the power quality problems in a hybrid power system connected to the grid [5]. Fuzzy and PI controllers were introduced for STATCOM, and their performances were studied over PV-Wind-Diesel- BESS based hybrid energy system [7]. The battery energy source (BESS) in conjunction with a static compensator was proposed [8] to mitigate power quality-related problems [9]. The performance analysis of a hybrid wind-diesel system with FPID and ANFIS-PID controller-based STATCOM was investigated [10]. However, no standard method is available for rule-based formation. Hence, the membership functions have to be tuned properly to minimize the error. Thus, to disregard the limitations of conventional and AI controllers, an AI controller called FIS controller is proposed in this work, as the FIS controllers are easy to be implemented [12-13]. Thus, in this study, a hybrid RES which comprises PV and wind system along with battery backup has been modeled. Here, the reactive power compensation is carried using fuzzy controlled STATCOM.

8.2 Solar PV System

8.2.1 System Description

The equivalent circuit PV cell (Figure 8.1) is made of semiconductor materials that convert incident solar light energy into electrical energy.

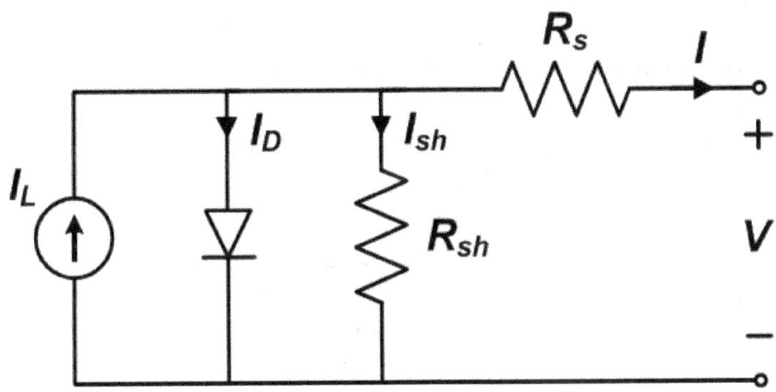

Figure 8.1 The simplified solar cell model.

The current /voltage characteristic of PV cell can be expressed as

$$I - I_L - I_0 \exp f(l) \left(\frac{(V+IR_S)}{nV_T} - 1\right) - \left(\frac{(V+IR_S)}{R_{Sh}}\right) \qquad (1)$$

where,

I and V are the current and voltage of solar panel,
$I_L = N_P I_L$ cell, and
N_P is the number of parallel cells,

Figure 8.2 depicts the IV/PV characteristics of the PV array utilized in this work.

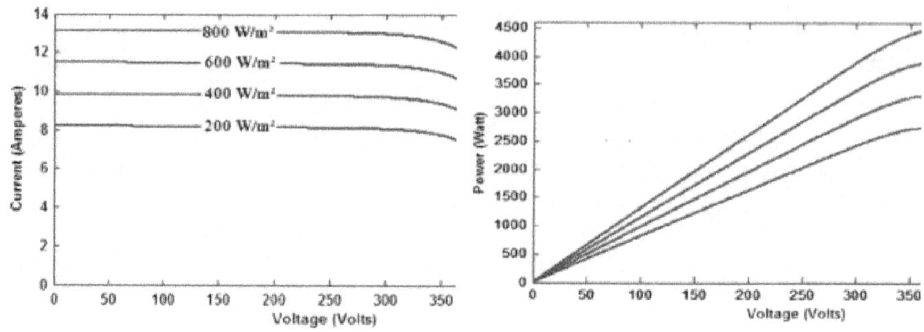

Figure 8.2 (a) IV and (b) PV characteristics of the proposed PV array.

It is found that at a lower irradiance value, a slight change in the maximum output voltage is observed, whereas output current changes drastically. Hence, there will be a drop in MPP.

8.2.2 Power Conditioning Unit Arrangement

The power-conditioning scheme comprises a boost converter with MPPT, a 3FVSI and an LC filter. The DC voltage acquired from the PV array is fed into the boost converter. Thus, to obtain the maximum power from the system, the P&O technique is employed. Figure 8.3 represents the flowchart of P&O topology. The boost converter implemented in this work is shown in Figure 8.4.

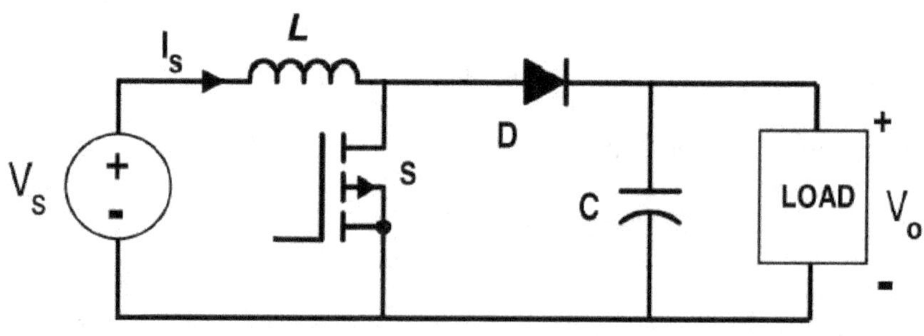

Figure 8.3 A typical boost converter.

Its operation is as follows:

During $(0 < t \leq ton)$: Switch remains „ON" and the voltage across L is V_i.

During (ton < t ≤ T): Switch remains „OFF" and the voltage across L becomes V_i-V_o.

The basic idea is to convert the kinetic energy available in the wind into mechanical energy, which rotates the turbine of an electrical generator to produce electrical power. The Bernoulli's equation gives us the power available from a wind turbine.

$$P_w = 1/2\ \rho AV^3 \tag{2}$$

$$P = 1/2\ C_P\ \rho AV^3 \tag{3}$$

where, ρ is the air density, V is the wind density, and A is the area swept by rotor respectively. However, practically power extracted from a wind turbine generator is the fraction of power *Pw*. The generator's rotor speed and pitch angle are the determining factors for the amount of power extracted (PT). Here, C_P is called power coefficient, whose value normally ranges around 0.593, generally referred to as Betz limit. The value of C_P can be controlled using two factors such as β and tip TSR. Finally, the output of the wind and PV systems is integrated and fed to VSI.

8.2.3 STATCOM

STATCOM is a compensating device that can be implemented to control the flow of reactive power in a network and thereby enhance a network's stability. The function of a capacitor in a STATCOM circuit is only to maintain dc voltage at a constant level to aid inverter operation. The capacitor does not produce any reactive power. The dc capacitor voltage can be controlled by having control over the phase angle difference between voltage source converter and line voltage. If the bus angle advances beyond the phase angle (α), the static compensator increases the injected voltage. The reactive power, thus, flows into the bus. Hence, in order to provide continuous voltage regulation in the microgrid, STATCOM should be controlled using the controller.

8.2.4 Design of Fuzzy Controller for STATCOM

The flow chart representation of the proposed fuzzy logic controller is established in Figure 8.4.

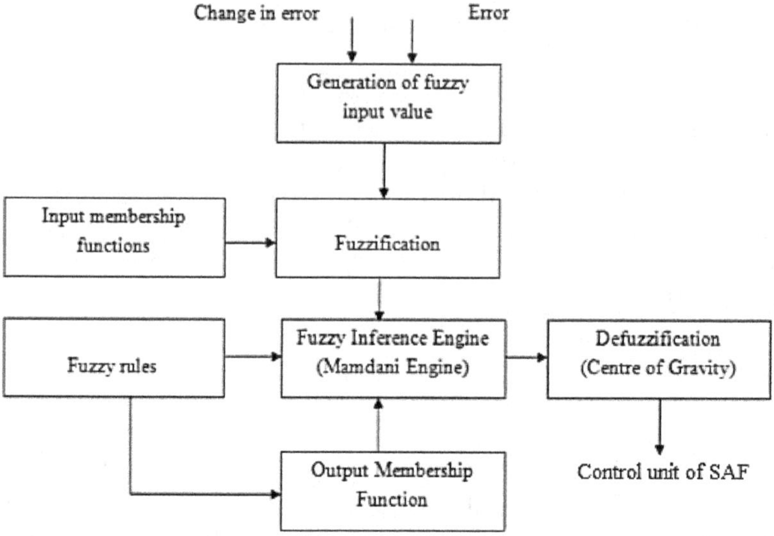

Figure 8.4 Flowchart representation of FLC.

The FLC utilized in this topology is characterized with
- Seven triangular MF for all inputs and outputs.
- continuous UoD.
- Implications are carried using Mamdani's „min" operator.
- Centroid method is adopted for defuzzification.

Figure 8.5 depicts MF adopted for both input and output.

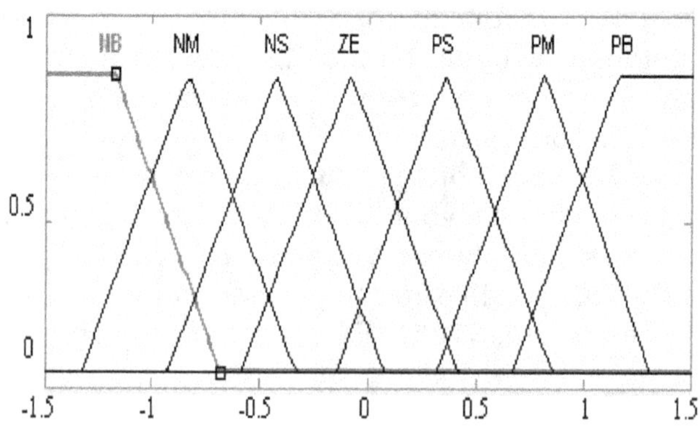

Figure 8.5a MF-e.

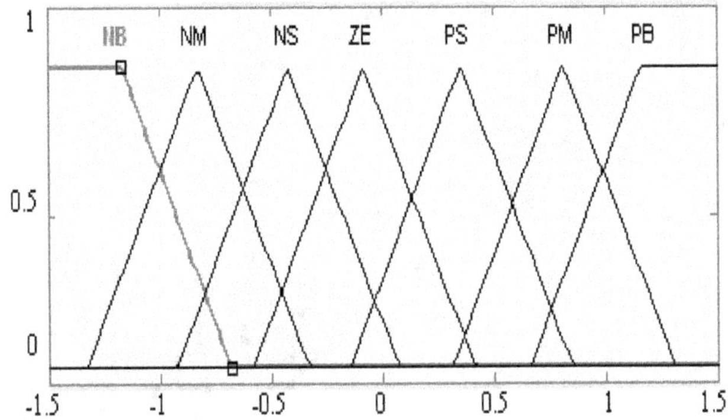

Figure 8.5b MF – Δe.

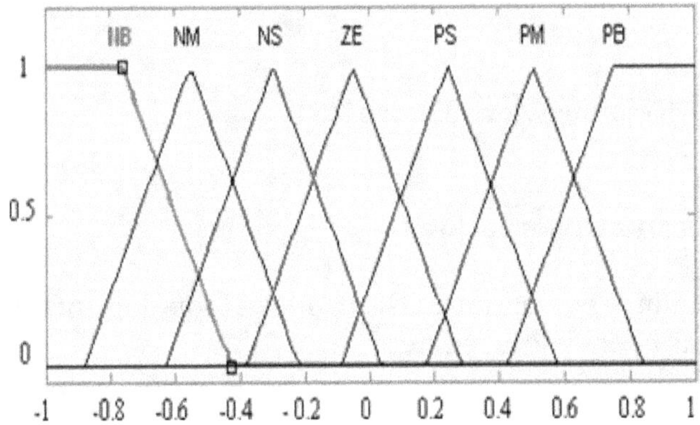

Figure 8.5c MF -Δu.

The 3D surface view of the rule base of FLC is shown in Figure 8.5d.

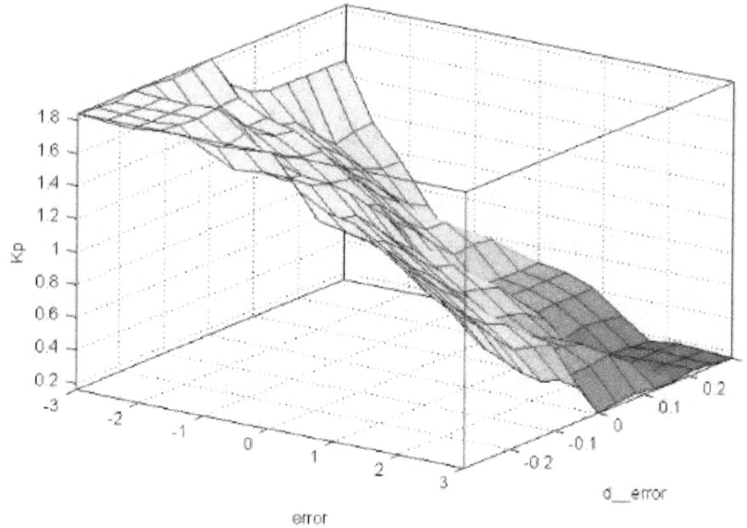

Figure 8.5d 3D surface view of proposed FLC.

8.3 Results and Observations

The performance evaluation of the proposed topology in the MATLAB platform has been discussed here.

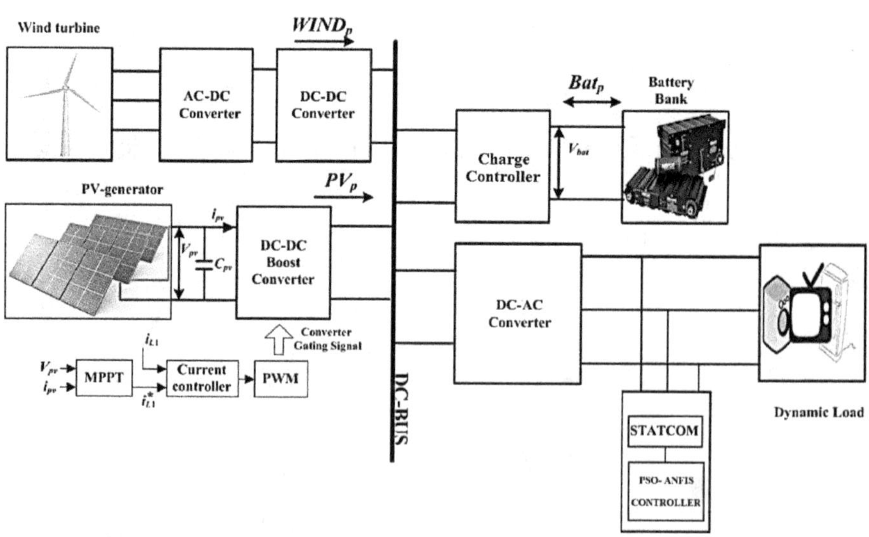

Figure 8.6 Topology of proposed methodology.

The proposed method's effectiveness is verified under different parameter analyses.

8.3.1 Improvement in Voltage Stability

The operation of microgrid without STATCOM s reflected in Figure 8.7.

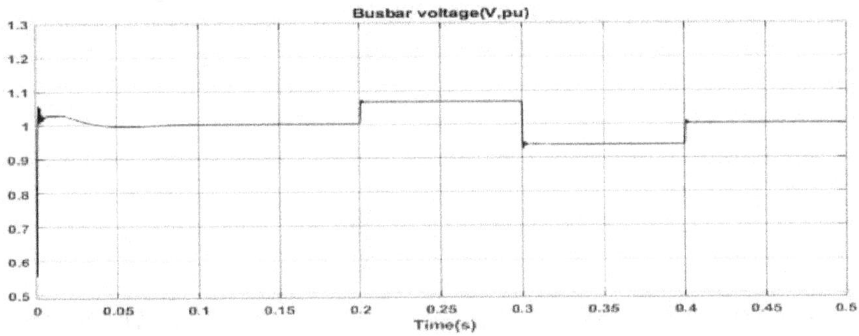

Figure 8.7 Bus bar voltage profile without STATCOM.

Thus, it is found that the voltage across the bus varies with time and results in fluctuation, which is about ±10%.

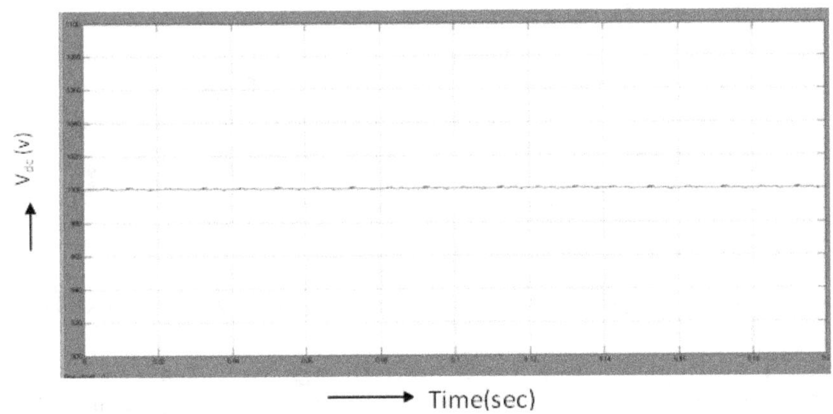

Figure 8.8 Bus bar voltage profile with STATCOM.

However, when the STATCOM is inserted, the voltage remains constant, as shown in Figure 8.8, and there is no fluctuation in voltage.

Hence, STATCOM with the proposed controller exhibits lower overshoot and minimized voltage fluctuation.

8.3.2 THD Calculation and FFT Analysis

Figure 8.9 describes the current and voltage waveforms of the system before STATCOM injection.

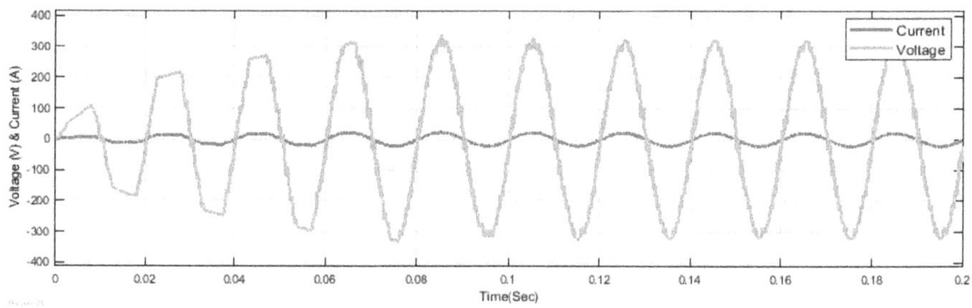

Figure 8.9 Voltage/current waveform of the system (without STATCOM).

Figure 8.9 clearly exhibits that there will be more non-linearity in voltage and current waveforms.

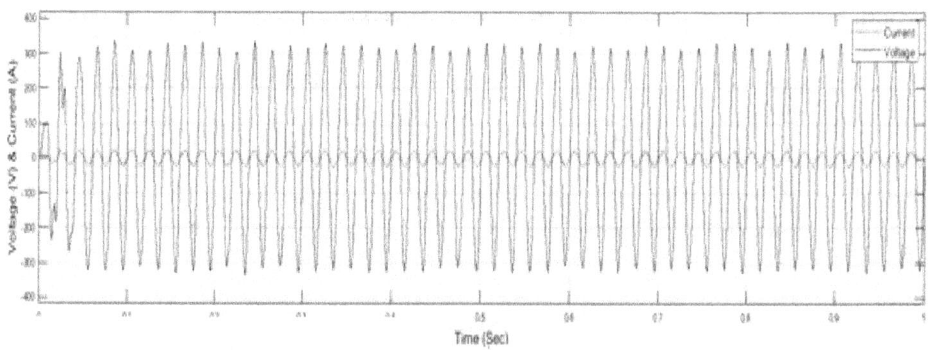

Figure 8.10 Voltage/current waveform of the system (with STATCOM).

Figure 8.10 depicts the current and voltage waveforms of FL-controlled STATCOM. It can be clearly understood that the static compensator removes the non-linearity to a great extent. Figures 8.11 and 8.12 represent the THD of the system obtained at the load terminal.

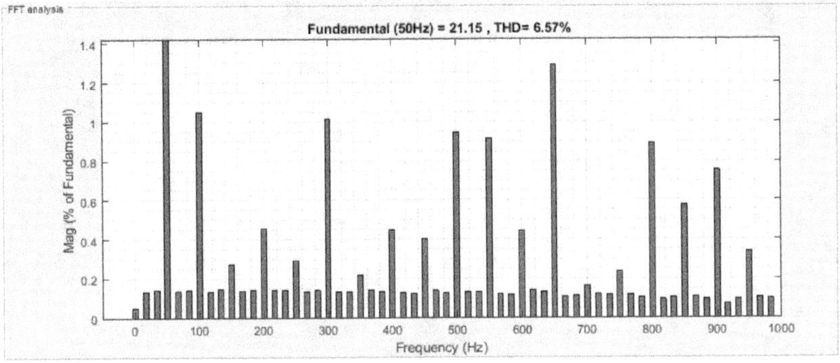

Figure 8.11 FFT analysis without STATCOM.

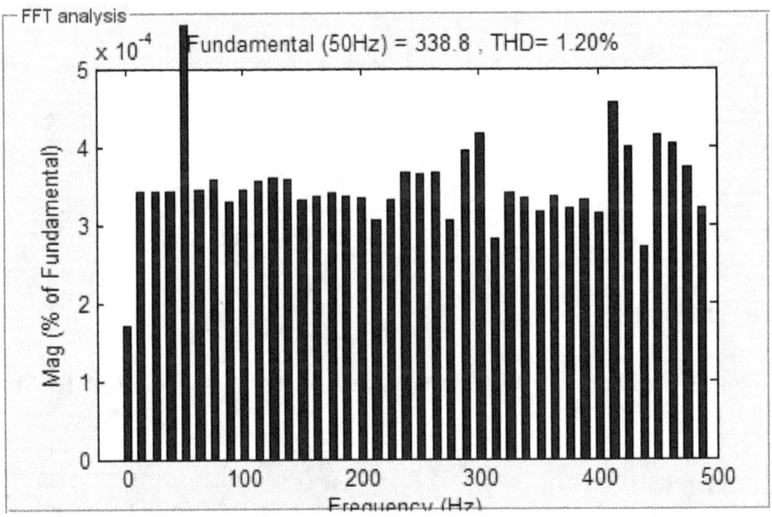

Figure 8.12 FFT analysis with proposed controller.

From the figures, the total harmonic distortion (THD) considering the maximum frequency as 500 Hz is found out to be 6.50% before compensation. After compensation, it is about 1.20% for the load voltage, i.e., well within the IEEE-519 prescribed limits.

8.3.2 Battery Performance

Here, the battery complements the part of the total power when the RES fails to supply the demand. Figure 8.13 portrays the SOC and output voltage of the battery.

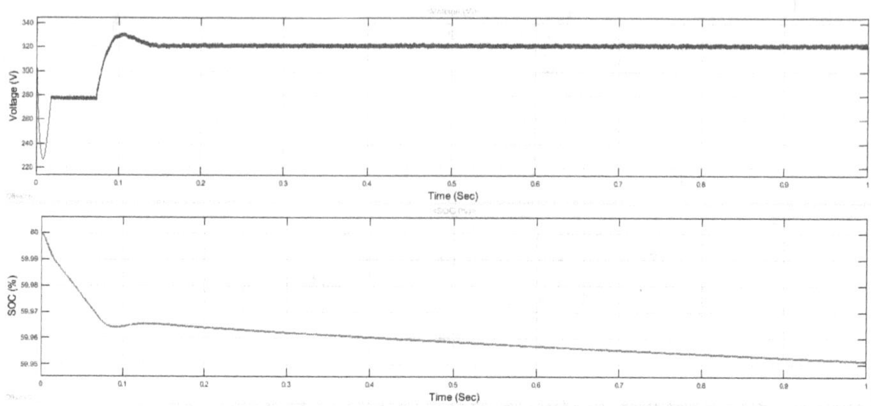

Figure 8.13 Output voltage/SOC of battery.

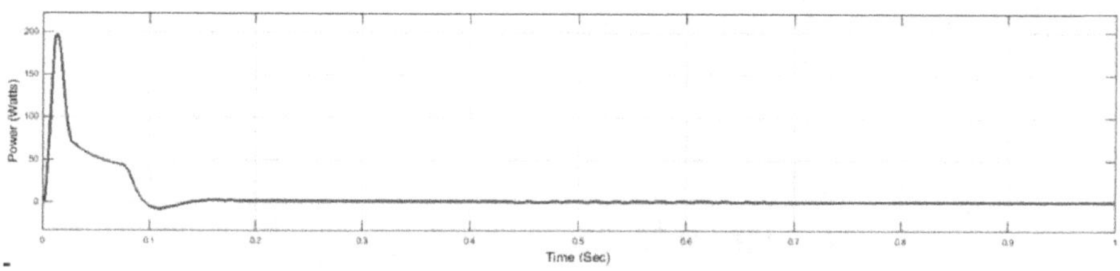

Figure 8.14 Battery power.

The power delivered by the proposed battery is depicted in Figure 8.14.

8.4 Conclusion

By understanding the relevance of solar tied microgrids in the current circumstances, authors have tried to explore and investigate findings and specific outcomes of the proposed scheme in the MATLAB environment. This work is principally focused on the PQ issues in solar PV based standalone or hybrid microgrid. In this work, the formulated MG comprises both PV and wind sources to supply power to the end-user. STATCOM is tailored to mitigate the power quality issues present in the MG. FLC is implemented to control the STATCOM's dc-link voltage at a constant level. With the help of this FLC-based STATCOM, the proposed work mitigates the harmonics present in the system along with reactive power compensation.

References

1. Wang, Q., Wang, B., Xu, W., & Xu, J. (2018, May). Research on STATCOM for reactive power flow control and voltage stability in microgrid. In 2018 13th IEEE Conference on Industrial Electronics and Applications (ICIEA) (pp. 2474-2479). IEEE.
2. Mohanty, K. B., &Pati, S. (2016, April). Voltage profile improvement of micro-grids using SMC based STATCOM. In 2016 Annual IEEE Systems Conference (SysCon) (pp. 1-8). IEEE.
3. Suja, K. R., & Raglend, I. J. (2012, March). Power quality improvement in grid connected wind energy system using STATCOM. In 2012 International Conference on Computing, Electronics and Electrical Technologies (ICCEET) (pp. 259-266). IEEE.
4. Chavan, P. M., &Chavan, G. P. (2017, August). Interfacing of hybrid power system to grid using statcom & power quality improvement. In 2017 International Conference on Information, Communication, Instrumentation and Control (ICICIC) (pp. 1-5). IEEE.
5. Mosobi, R. W., Chichi, T., &Gao, S. (2015). Power quality analysis of hybrid renewable energy system. Cogent Engineering, 2(1), 1005000.
6. Mohod, S. W., & Aware, M. V. (2010). A STATCOM-control scheme for grid connected wind energy system for power quality improvement. IEEE systems journal, 4(3), 346- 352.
7. Suvire, G. O., & Mercado, P. E. (2009, September). Improvement of power quality in wind energy applications using a DSTATCOM coupled with a flywheel energy storage system. In 2009 Brazilian Power Electronics Conference (pp. 58-64). IEEE.
8. Anulal, A. M., Mohan, A., & Lathika, B. S. (2015, November). Reactive power compensation of wind-diesel hybrid system using STATCOM with Fuzzy tuned and ANFIS
9. Das T.K., Banik A., Chattopadhyay S., Das A. (2019) Sub-harmonics Based String Fault Assessment in Solar PV Arrays. In: Chattopadhyay S., Roy T., Sengupta S., Berger-Vachon C. (eds) Modelling and Simulation in Science, Technology and Engineering Mathematics. MS-17 2017. Advances in Intelligent Systems and Computing, vol 749. Springer, Cham.
10. T. K. Das, A. Banik, S. Chattopadhyay and A. Das, "FFT based Classification of Solar Photo Voltaic Microgrid System," 2019 Second International Conference on Advanced Computational and Communication Paradigms (ICACCP), Gangtok, India, 2019, pp. 1-5,
11. A. Banik and A. Sengupta, (2021) "Scope, Challenges, Opportunities and Future Goal Assessment of Floating Solar Park," 2021 Innovations in Energy Management and Renewable Resources (52042), Kolkata, India, pp. 1-5.
12. A. Banik, A. Shrivastava, R. Manohar Potdar et al., (2021) Design, Modelling, and Analysis of Novel Solar PV System using MATLAB, Materials Today: Proceedings, https://doi.org/10.1016/j.matpr.2021.06.226

13. Das T.K., Banik A., Chattopadhyay S., Das A. (2021) Energy-Efficient Cooling Scheme of Power Transformer: An Innovative Approach Using Solar and Waste Heat Energy Technology. In: Ghosh S.K., Ghosh K., Das S., Dan P.K., Kundu A. (eds) Advances in Thermal Engineering, Manufacturing, and Production Management. ICTEMA 2020. Lecture Notes in Mechanical Engineering. Springer, Singapore. https://doi.org/10.1007/978-981-16-2347-9_17

www.ingramcontent.com/pod-product-compliance
Lightning Source LLC
LaVergne TN
LVHW011520200926
848484LV00001B/92

* 9 7 8 1 9 2 2 6 1 7 2 8 6 *